IFIP Advances in Information and Communication Technology

719

IFIP Advances in Information and Communication Technology

The IFIP AICT series publishes state-of-the-art results in the sciences and technologies of information and communication. The scope of the series includes: foundations of computer science; software theory and practice; education; computer applications in technology; communication systems; systems modeling and optimization; information systems; ICT and society; computer systems technology; security and protection in information processing systems; artificial intelligence; and human-computer interaction.

Edited volumes and proceedings of refereed international conferences in computer science and interdisciplinary fields are featured. These results often precede journal publication and represent the most current research.

The principal aim of the IFIP AICT series is to encourage education and the dissemination and exchange of information about all aspects of computing.

More information about this series at https://link.springer.com/bookseries/6102

Robert M. Davison · David Kreps
Editors

Human Choice and Computers

16th IFIP International Conference
on Human Choice and Computers, HCC 2024
Phuket, Thailand, September 8–10, 2024
Proceedings

 Springer

Editors
Robert M. Davison (iD)
City University of Hong Kong
Kowloon, Hong Kong

David Kreps (iD)
University of Galway
Galway, Ireland

ISSN 1868-4238   ISSN 1868-422X (electronic)
IFIP Advances in Information and Communication Technology
ISBN 978-3-031-67534-8   ISBN 978-3-031-67535-5 (eBook)
https://doi.org/10.1007/978-3-031-67535-5

This Springer imprint is published by the registered company Springer Nature Switzerland AG
The registered company address is: Gewerbestrasse 11, 6330 Cham, Switzerland

If disposing of this product, please recycle the paper.

Preface

This book contains the proceedings of the 16th International Conference on Human Choice and Computers (HCC 2024), which was held at the Dusit Thani Laguna, Phuket, Thailand, during September 8th to 10th, 2024. The conference was organized by the International Federation for Information Processing (IFIP) Technical Committee 9 (TC9): Information and Communication Technology (ICT) and Society.

The 17 submitted papers were double-blind reviewed by three members of the Programme Committee, during a six-week review period. Nine full papers were accepted, based on review scores. Papers submitted by a programme committee member were managed by the programme chairs independently.

The conference co-chairs, David Kreps (chair of IFIP's Technical Assembly) and Robert Davison (chair of TC9), chose the theme for this year's conference: "Humans, Technological Innovations and Artificial Intelligence: Opportunities and Consequences". Tracks were advertised in the call for papers addressing a range of concerns across the working groups of TC9. We are most grateful to the following colleagues who undertook reviews of one or more papers: Angsana Techatassanasoontorn, Antonio Diaz Andrade, Chris Leslie, Christophe Viguerie, Evelyn Ng, Federico Iannacci, Hideyuki Nakashima, Hugo Lotriet, Judy van Biljon, Julie Cameron, Marco Marabelli, Minna Rantanen, Miroslaw Sikora, Norberto Patrignani, Petros Chamakiotis, Richard Taylor, Sam Takavarasha, Takayuki Matsuo and Willem van Eekelen.

The papers selected for this book are based on both academic research and the professional experience of information systems practitioners working in the field. It is the continued intention of TC9 that academics, practitioners, governments and international organizations alike will benefit from the contributions of these proceedings. Details of IFIP TC9's activities are posted at http://www.ifiptc9.net/.

July 2024   Robert M. Davison
                                                                                         David Kreps

Organization

Conference Chairs

David Kreps   University of Galway, Ireland
Robert M. Davison   City University of Hong Kong, Hong Kong

Programme Committee Chair

Endrit Kromidha   University of Birmingham, UK

Organizing Committee Chair

Siripan Deesilatham   University of the Thai Chamber of Commerce,
                                   Thailand

Proceedings Editor

Robert M. Davison   City University of Hong Kong, Hong Kong

Programme Committee

Angsana   Auckland University of Technology, New Zealand
  Techatassanasoontorn
Antonio Díaz Andrade   University of Agder, Norway
Chris Leslie   Prince of Songkla University, Thailand
Christophe Viguerie   City University of Hong Kong, Hong Kong
Evelyn Ng   University of Sydney, Australia
Federico Iannacci   University of Sussex, UK
Hideyuki Nakashima   Hakodate University, Japan
Hugo Lotriet   University of South Africa, South Africa
Judy van Biljon   University of South Africa, South Africa
Julie Cameron   Info T.EC Solutions Pty Ltd, Australia
Marco Marabelli   Bentley University, USA
Minna Rantanen   University of Turku, Finland
Miroslaw Sikora   University of Silesia in Katowice, Poland
Norberto Patrignani   Politecnico di Torino, Italy
Petros Chamakiotis   ESCP Madrid, Spain
Richard Taylor   IBO, UK
Sam Takavarasha   Women's University in Africa, Zimbabwe
Takayuki Matsuo   Harvard Law, USA
Willem van Eekelen   Independent Consultant, UK

Humans, Technological Innovations and Artificial Intelligence: Opportunities and Consequences

Robert M. Davison[1] [iD] and David Kreps[2] [iD]

[1] City University of Hong Kong
isrobert@cityu.edu.hk
[2] University of Galway, Ireland
david.kreps@universityofgalway.ie

1 Choice

Fifty years ago, the first Human Choice and Computers Conference (HCC) was organized, chaired by IFIP's President Heinz Zemanek, with the proceedings edited by Mumford and Sackman. An overarching theme of that first conference was a concern over the way the participants felt people were being forced to use computers in dehumanizing ways. They argued that sociotechnical problems must be solved in ways that foreground the interests of workers and communities and that, ultimately, human needs must take precedence over technological and economic considerations. Those concerns have never really left us and indeed are close to the theme of the 16th HCC.

Over the last two years, we have seen the meteoric rise of Generative Artificial Intelligence (generally known as GAI) tools like ChatGPT, ChatSonic, LaMDA, Neeva AI, DragonFly, etc. As we claim to move towards a more human-centric Industry 5.0, these, and other, technological innovations are challenging many of the existing relationships and choices that exist between humans and computers. The challenges have been documented extensively in the popular press, but also in academia. The challenges exist at the juncture of human needs on the one hand, and technological and economic considerations on the other. While the technology may have changed, the events of 50 years ago still seem very fresh.

Some scholars and pundits are profoundly negative in their evaluations of AI technologies, suggesting that these tools will upend many aspects of the status quo in any domain where human creativity dominates, notably education, journalism, research, governance, and of course crime. Others, perhaps those with a Machiavellian inclination, are quick to see the advantages associated with the new technology and argue that developments and innovations of this kind cannot simply be stopped by fiat. Indeed, although they may have the potential to eliminate creative work, they are themselves the products of creative and fertile imaginations. Unsurprisingly, new tools (themselves premised on AI) that are claimed to detect AI-created materials have also emerged, perhaps initiating a 'war' between the two sides.

What we can expect is that just as the new technology may solve some problems, it may exacerbate others. For instance, as we noted in the call for papers for the previous HCC15 conference, the encroaching influence of machine learning based systems that can embed the biases inherent in the data they have learnt from, threatens to entrench the societal problems of the past, rather than redress them. A wide range of ethical issues are certainly associated with the new technology, which we suggest will prove to offer a cornucopia of new research opportunities.

2   Papers

The papers accepted for this conference fell broadly into two categories: those which dealt with a more general selection of issues related to AI, including historical, legal and methodological perspectives; those which consider a more contextual analysis of issues, with cases of particular situations in Indonesia, France, Thailand and the Global South.

We start with two papers that document historical perspectives into AI. Sugimoto explores the historical (postwar) context of AI and robotics development in Japan. The historical context is pertinent because unlike in other countries where AI is both funded by and harnessed for the military-defence establishment, Japan's post-war constitution severely restricts military-defence spending. This unique set of circumstances has led to an AI development paradigm that is markedly different. In contrast to this Japanese perspective, Hongladarom, Sakprasert and van der Vaeren offer a review of AI research in Thailand, drawing on interview data with Professor Asanee Kawtrakul, one of the pioneering founders of AI in the country. Both historical trends, much of which involved Natural Language Processing for the Thai language, and future developments are identified.

To continue with the AI theme, Kromidha and Davison explore the opportunities for AI-augmented decision making. They note ethical biases often associated with AI, for instance algorithmic hallucinations and a lack of transparency, and report on their experimental investigation into ChatGPT's responses to business questions that are typically encountered in practice.

Nakashima reports on new legislative issues that crop up in the case of app-stores in Europe, Japan and the US. Their legal research reminds us that similar developments are likely to follow in the AI-space; even though laws frequently lag technological innovations by a number of years, we can anticipate that the laws will emerge and challenge the AI innovations that we are currently seeing.

Technological innovations can also work within existing legal frameworks. Van Vuuren and Manala explain how a machine-learning phishing detection model can serve the e-banking environment. Meanwhile, a prominent dark side to these techno-logical innovations surfaces in the form of online gambling: van Eekelen investigates this phenomenon which constitutes an invisible (and perhaps inevitable) curse for the Global South.

Turning to country level studies, Er, Davison, Khoir and Fauzia report on a multi-case study investigation into the digital transformation (DT) of warungs in Indonesia. Such nano-enterprises (warungs are small family stores) are seldom the focus of DT research, yet also stand to profit from this contemporary trend if they can successfully bridge the human-technology gaps that are common in the Global South. Returning to AI, Desbois explores how AI creates value for the French health sector. Special considerations must be paid to the European regulatory context of the single digital market and personal data protection regulations.

Finally, we conclude the papers with a topic that runs throughout the themes of this conference: digital inequality. Khan, Haque and Shahriar discuss the development of survey instruments that can be used to measure digital equality at the individual level.

First, however, to begin the proceedings, our two keynote speakers kindly provided summaries of their presentations for us to reproduce in this book. In the first keynote, Prof Steffen Kromer shared a transnational perspective on AI cultural challenges, noting that the current US-China dominance directly implies the exclusion of other cultural perspectives. He argued that in order to preserve cultural identities worldwide, a critical and transnational approach to AI development is needed. In the second keynote, Apivadee Piyatumrong explored the landscape of AI governance in Thailand with a particular focus on ethics and trust issues associated with improving the quality of life of people in Thailand, in conjunction with Thailand's national AI strategy and action plan (2022–2027).

We hope and trust you will enjoy reading the texts within these pages.

Contents

A Transnational Perspective on Artificial Intelligence Cultural Challenges

Steffen Kromer[(⊠)]

Technik Und Kultur, Victoria, International Hochschule Für Wirtschaft, Bernburger Str. 24-25, 10963 Berlin, Germany
steffen.kromer@victoria-hochschule.de

Abstract. US and Chinese multinational corporations are actively leading and shaping artificial intelligence strategies. Such dominance from these two countries requires a more transnational perspective on artificial intelligence (AI) and its implications on cultures worldwide. Exclusion of other cultures has important implications on national AI strategies of major players and other countries. With a view to preserve cultural identities worldwide, a critical approach and transnational perspective on AI cultural challenges is needed.

1 Introduction

This paper calls for a renewed examination of artificial intelligence (AI) from an interdisciplinary perspective, focusing particularly on societal aspects such as cultural challenges.

Intelligent machines, smart devices, and autonomous decision-making systems are increasingly integral to AI [1]. However, AI, a branch of computer science [2], originated from the idea that all aspects of learning or intelligence could be precisely described and simulated by a machine [3]. As a result, AI has rapidly developed and now permeates every aspect of modern society [4]. AI's focus on abstraction, logical formalism, programming, and algorithmic details often prioritizes synthesis over analysis and engineering over knowledge [2]. Thus, the critique [5] that current AI systems fail to link symbols with their meanings should be taken seriously. For more insights on these attributes, see Nilsson [2].

Initially, AI was envisioned through fantasies, possibilities, demonstrations, and promises [6]. Over the past 50 years, it has transformed significantly, now impacting daily life globally [4]. A vast body of literature has emerged on AI [1, 6–15], covering critical areas e.g. healthcare [16] and humanitarian aid, as well as everyday activities like dating [4], autonomous driving, translation services, oil drilling, and even milking cows [17]. Additionally, AI has seamlessly integrated into our daily lives, enhancing knowledge and capabilities in transportation, traffic avoidance, social connections, movie recommendations, and healthier food choices [18].

Keynote Talks. Our two keynote speakers have kindly provided summaries of their presentations for the proceedings.

© IFIP International Federation for Information Processing 2024
Published by Springer Nature Switzerland AG 2024
R. M. Davison and D. Kreps (Eds.): HCC 2024, IFIP AICT 719, pp. 1–8, 2024.
https://doi.org/10.1007/978-3-031-67535-5_1

In the twenty-first century's "man meets machine" reality [18], AI, including embodied AI in robotics and machine learning, may boost economic and social welfare [4]. However, research indicates that AI also introduces cultural, ethical, social, and legal challenges related to data accessibility and integrity, privacy, safety, algorithmic bias, transparency, and interpretation of outcomes [19, 20].

Therefore, AI might be misused or behave unpredictably and harmfully [4]. As Minsky [11] states, "a computer can do, in a sense, only what it is told to do". Addressing cultural challenges from a transnational perspective of AI is increasingly important because, as Floridi [21] notes, "the digital revolution transforms our views about values and priorities, good behaviour, and what sort of innovation is not only sustainable but socially preferable—and governing all this has now become the fundamental issue".

While AI has the potential to improve lives globally, this paper argues from a transnational perspective that integrating AI into broader society raises cultural challenges. Most AI applications, which use learning techniques from big data sets, act on predictions derived from statistical patterns [5]. These predictions influence many aspects of contemporary life, making it essential to consider cultural challenges.

Given AI's profound impact on people's lives and the world, this paper highlights some of the often-overlooked cultural challenges associated with AI's rise. Considering AI's global effects, cultural challenges must be seriously addressed in terms of transparency, fairness, and linking symbols to their specific cultural meanings. Therefore, the author calls for a multidisciplinary and transnational perspective.

The next section situates this paper within the existing literature. It provides a brief overview of the current global state of AI development and presents prominent examples emphasizing the need to consider cultural challenges related to AI. Finally, the author advocates for a more inclusive, and thus more interdisciplinary, approach that keeps AI connected to the transnational human context in which it operates in.

2   Background

Academics and regulators are struggling to keep pace with the flood of articles, principles, regulatory measures, and technical standards related to AI governance [4]. In the first half of 2018 alone, over a dozen countries unveiled their latest AI strategies [4]. Additionally, industries have been formulating their own AI principles and launching initiatives to establish best practices. They have also been actively involved in creating regulations for AI through direct engagement and lobbying [4]. Currently, the leading nations in AI are the United States and China. In the US, the dynamism of major technology corporations (such as Google, Amazon, Facebook, Apple, Microsoft, and IBM), extensive university research, and access to private capital have positioned the US as a pioneer in AI [8]. However, the initial advancements credited to the US are gradually being overshadowed by the rapid progress of China's tech industry and strategic government initiatives. China's innovative technology giants like Baidu, Alibaba, and Tencent, supported by a vast network of research labs and access to private and government funds, mirror the roles of Google, Amazon, and Facebook [8, 22].

In 2017, the Chinese government announced its plan to become the leading global centre for AI by 2030, aiming to create a US$150 million market focusing on areas such as surveillance, transportation, defence, and healthcare. Conversely, the European Union, especially the United Kingdom (despite Brexit) and France, boasts crucial research centres, numerous AI start-ups, and robust policies aimed at reducing the dominance of US and Chinese digital markets. However, Europe has yet to produce digital multinationals comparable to those in the US and China, partly due to lower access to private capital [8]. Other active AI countries include Israel, which has 40 times more AI startups per capita than the US, as well as Canada, Japan, India, Mexico, and South Korea [4, 8].

"Success in creating AI would be the biggest event in human history. Unfortunately, it might also be the last, unless we learn how to avoid the risks" [16, 23]. "Leveraging opportunities and tackling the challenges posed by this technology… Will transform society and industry as profoundly as electricity did" [8]. In the current global AI development landscape, data has become the raw material of production and a new source of immense economic and social value and power [24]. Therefore, the collection and analysis of data increasingly shape the production of goods and services to maximize profits [25–27]. Numerous, scholars interpret the progress and deployment of AI between China and the US through a rivalry lens. Thus, countries that do not invest in AI and lack their own strategies will arguably be compelled to consume products and services from predominantly US and Chinese tech multinationals.

Consequently, if cultural aspects and their associated challenges, including linking symbols to their meanings in their full richness and diversity, are not explicitly incorporated into the global AI agenda, the likelihood of bias re-emerging is almost certain. Moreover, these challenges concern not only the functioning and data ownership of AI systems but also how AI systems are perceived and conceptualized, the cultural challenges involved, and their potential consequences. Despite being outside the scope of this paper, cultural challenges, related to governing AI, shall ideally consider all national cultures, nations, as well as the many variations of other cultures, since AI's impact is a truly global phenomenon today [28]. Therefore, calling for more work, scrutinising cultural challenges in the field of AI and its development is of global importance.

3 Cultural Challenges of Artificial Intelligence

The following paragraphs highlight significant concerns related to AI governance and call for closer examination of the cultural challenges arising from the industry's active role in shaping AI in nearly all its aspects. Cultural challenges are becoming increasingly important in discussions about AI. Additionally, automated applications are becoming more prevalent in daily life, raising the risk that the way these systems are engineered could lead to decisions that negatively impact societies globally. Indeed, the design of algorithms and the selection of input data can hide biases related to gender, race, and other factors, thereby amplifying the prejudices of those who created the applications [5].

This could result in discrimination against certain groups or selective censorship of content. For example, an international beauty contest in 2016 evaluated participants based on AI and revealed a bias: out of over 6,000 participants from 100 countries, 44 of the top-listed winners were Caucasian, and only one was a person of colour. This indicates that the training data did not sufficiently include people of colour or Asians, leading to bias against marginalized or ethnic groups [8].

Another example is the reinforcement of gender stereotypes through algorithms, particularly in AI-based translation systems that use machine learning. For instance, the Turkish phrase 'O bir bilim adami' (he/she is a scientist) is translated by Google Translate as 'he is a scientist,' while 'O bir hemsire' (he/she is a nurse) is translated as 'she is a nurse,' reflecting gender assumptions embedded in the design.

Gender biases are also evident in search engines like Google and Baidu. Searching for 'successful person' primarily returns images of Caucasian men, with few women or people of other ethnicities [8], showing that AI systems have not effectively implemented the task of assigning meaning to symbols.

Discrimination and censorship risks arise not only from how AI systems are designed, and the data sets used but also from how founding technology multinationals integrate them into the user experience. Virtual assistants often have female names, such as Cortana, Alexa, and Siri [8]. These choices are not driven by algorithms, data, or programmers but by gender stereotypes prevalent in the industry, suggesting that women are still viewed primarily as assistants.

What is missing is the inclusion of a cultural perspective. Despite extensive literature on the ethics of AI [4, 7, 29], studies often overlook cultural variables, rendering directives as mere declarations of intent without specific applications in a diverse world. Questions arise: What does 'socially beneficial AI' mean? According to whose cultural values? For which groups? Under what conditions? And at what time? Can a phenomenon be considered beneficial if it simplifies consumers' lives in the short term but drastically transforms or eradicates cultures in the long term? Who defines social benefit, and on what basis are AI algorithms designed?

Similarly, Lipton and Steinhardt [30] argue that complex social concepts such as culture, gender, race, and appearance should not be simplified to mere statistics in algorithm design.

As previously mentioned, many industry leaders in AI are based in the US and China. This raises concerns about how much AI systems reflect US and Chinese societies and the preferences of their technology multinationals. AI programming primarily relies on data sets, with most technical innovations controlled by a few US and Chinese companies [4, 8]. Given these companies' prominent roles in regulatory initiatives, it is crucial to prevent their agendas from solely driving global AI development, which may not align with other cultural contexts, such as differing EU privacy regulations.

Thus, the above depicted circumstances delineate a challenge, which in essence makes plain that to collaborate globally, there is a clear need for cultural sensitivity, which is context specific, hence different approaches are needed. Therefore, Acharya's [31] claim, that it is crucial to avoid situations, in which one or two countries try to impose their cultural values, symbols and norms on the rest of the world gains great importance. Therefore, different geographic and cultural regions call for different governance approaches [32].

Nevertheless, AI systems are often seen as 'black boxes' [33], complex and difficult to explain [34]. Scholars such as Kroll [35] counter that algorithms are understandable, arguing that problematic systems should be viewed as malpractice by their controllers rather than inscrutable technology. However, the perception of complex, inscrutable technology often justifies the close involvement of the AI industry in policymaking and regulation. Industry stakeholders involved in these processes, at least in the US, are often the same elite group leading online marketing and data collection. Companies like Google, Facebook, Amazon, Apple, IBM, Baidu, Alibaba, and Tencent can gather large amounts of data to enhance innovative AI services [4].

This AI development consolidates the market power of big companies and validates their role in governance processes. The European Commission recently appointed a High-Level Expert Group on AI [4] to work on implementing the European AI strategy. The group's 52 members come from diverse backgrounds, but nearly 50% are still from the industry, 17 are from academia, and 4 from civil society. Academics argue that civil society, often most affected by AI systems, should have a fair role in contributions [36].

Thus, one of the key challenges from an interdisciplinary perspective is to identify the areas where international agreement in terms of negotiated cultural variations is most important and separate these from areas where it is more appropriate to respect a plurality of approaches. Ultimately, it seems that mistrust between cultures and dissimilar geographical regions is one of the biggest contributors, that fuels cultural challenges, as policymakers, technologists and scholars in China and the US are highly subjected to it; due to political and historical tensions, as well as diverging philosophical traditions, which actively shape the culture of these regions [20].

Nevertheless, it would be too simple to assume that the main cause of cultural challenges within the arena of AI is that China and the US rest upon fundamentally different values, norms, and symbols, causing them to produce dissimilar, perhaps conflicting, perceptions of how AI, and its related technologies should be utilised, engineered, and governed [37]. As a result, the notion that 'Western' and 'Eastern' cultures are fundamentally opposed to each other severely oversimplifies and diminishes the sheer variety of these geographical regions. Even within East Asia, particularly China, Japan and Korea significant cultural differences prevail. Similarly, the cultures and philosophies representing the 'West' are not all the same [38, 39]. However, cultures, regardless of place and time, are arguably ever-changing entities that constantly evolve [39].

The somewhat dangerous oversimplification of the terms 'Western' and 'Eastern' becomes apparent in Johnston & Shen's [40] work, which demonstrates that misperceptions exist in both contexts. The diversity of and in Western societies is often oversimplified and narrowed down to a single pattern of Northern American life in China. Likewise, China and Chinese culture are not fully understood, and misperceptions remain. Taking privacy as a cultural example, one cannot simply assume that privacy is valued more in 'Western' cultures opposed to 'Eastern' cultures, however, we need a more nuanced understanding of cultural symbols, values, and norms [41]. Thus, understanding the different connotations and cultural assumptions underpinning cultural values, norms and symbols in different context is important in the context of AI.

Therefore, the concerns discussed regarding cultural, and, to a lesser extent, ethical challenges highlight the need to address cultural variables in future AI development.

4 Concluding Remarks

The argument presented in this paper should not be seen as dismissing the efforts made in the field of AI. There is much to be learned from the ongoing discourse, but long-term transnational benefits will only be realized if the concerns raised here are addressed comprehensively, since AI is a truly global matter. Therefore, it is essential to critically examine the underlying goals of AI, along with the associated cultural challenges and impacts, especially regarding industry-driven development on the global stage, considering the dominant roles of the US and China.

Ultimately, as data has become a key resource and a significant source of economic and social value, important questions must be asked, such as: What does 'socially beneficial AI' mean? According to which cultural values? For which groups? Under what conditions? And at what time? This paper emphasizes the importance of recognizing the cultural aspects and its challenges in all their complexity when addressing AI. Furthermore, more initiatives led by countries other than the US are needed, like the EU's approach, and the rise of Chinese technology multinationals must be taken seriously. Other scholars also stress the need to move beyond cultural and ethical discussions to consider additional core values [21, 36, 42]. The ongoing mistrust between China on one hand and the US on the other, places a particular strain on this. As previously mentioned, overcoming all cultural challenges related to AI is beyond this paper's scope, however, making efforts to achieve greater understanding between the current 'AI powerhouses' may help tremendously in promoting greater diversity of cultural representations worldwide within the field of AI.

Overall, the transnational perspective advocated in this paper highlights the intricate considerations of AI and calls for a more inclusive, interdisciplinary approach. According to Kroll [35], "in general, opacity in socio-technical systems results from power dynamics between actors that exist independent of the technical tools in use. No artifact is properly comprehended without reference to its human context, and software systems are no different".

Moreover, in a technological landscape primarily dominated by the US and China, and to a lesser extent by Europe, Israel, Canada, Japan, and South Korea, there is a real risk of creating a dual divide, leading to the exclusion of diverse cultures in national AI strategies. This could result in countries losing their cultural identities, which may have severe consequences.

This paper has provided insights into a more transnational perspective on AI and its cultural challenges, highlighting the need for further research into practical solutions for these challenges. Over time, more detailed implications of cultural challenges for AI may become clearer.

References

1. Gunkel, D.J.: Communication and artificial intelligence: opportunities and challenges for the 21st century. Futures Commun. 1(1), 1–25 (2012)
2. Nilsson, N.: Principles of Artificial Intelligence. Tioga Press, Menlo Park, CA (1980)

3. McCarthy, J., Minsky, M.L., Rochester, N., Shannon, C.E.: A proposal for the Dartmouth summer research project on artificial intelligence, August 31, 1955. AI Mag. **27**(4), 12–14 (2006)
4. Cath, C.: Governing artificial intelligence: ethical, legal and technical opportunities and challenges. Philos. Trans. R. Soc. A: Math. Phys. Eng. Sci. **376**(2133), 20180080 (2018)
5. Lu, H., Li, Y., Chen, M., Kim, H., Serikawa, S.: Brain intelligence: go beyond artificial intelligence. Mob. Netw. Appl. **23**, 368–375 (2017)
6. Buchanan, B.G.: A (very) brief history of artificial intelligence. AI Mag. **26**(4), 53–60 (2005)
7. Etzioni, A., Etzioni, O.: Incorporating ethics into artificial intelligence. J. Ethics **21**(4), 403–418 (2017)
8. Kulesz, O.: Culture, platforms and machines: The impact of artificial intelligence on the diversity of cultural expressions (Report DCE/18/12.IGC/INF.4). Paris, France: United Nations Educational, Scientific and Cultural Organization (2018)
9. McCarthy, J.: Epistemological problems of artificial intelligence. In Webber, B.L., Nilsson, N.J. (Eds.), Readings in artificial intelligence, Morgan Kaufmann Publishers, pp. 459–465 (1977)
10. McCarthy, J., Hayes, P.J.: Some philosophical problems from the standpoint of artificial intelligence. In: Webber, B.L., Nilsson, N.J. (Eds.), Readings in artificial intelligence, Morgan Kaufmann Publishers, pp. 451–458 (1969)
11. Minsky, M.: Steps toward artificial intelligence. Proc. IRE **49**(1), 8–30 (1960)
12. Müller, V.C., Bostrom, N.: Future progress in artificial intelligence: a poll among experts. AI Matters **1**(1), 9–11 (2014)
13. Müller, V.C., Bostrom, N.: Future Progress in Artificial Intelligence: A Survey of Expert Opinion. In: Müller, V.C. (ed.) Fundamental issues of artificial intelligence, pp. 553–571. Springer, Cham (2016)
14. Scherer, M.U.: Regulating artificial intelligence systems: risks, challenges, competencies, and strategies. Harvard J. Law Technol. **29**(2), 354–398 (2016)
15. Makridakis, S.: The forthcoming artificial intelligence (AI) revolution: its impact on society and firms. Futures **90**, 46–60 (2017)
16. Xia, J., Xie, F., Zhang, Y., Caulfield, C.: Artificial intelligence and data mining: algorithms and applications. Abstr. Appl. Anal. **2013**, 948 (2013)
17. Russel, S.J., Norvig, P.: Artificial Intelligence: A Modern Approach. Pearson Education Limited, Malaysia (2016)
18. Perc, M., Ozer, M., Hojnik, J.: Social and juristic challenges of artificial intelligence. Palgrave Commun. **5**, 61 (2019)
19. Przegalinska, A.: State of the art and future of artificial intelligence. Policy Briefing, Policy Department for Economic, Scientific and Quality of Life Policies. Directorate General for Internal Policies. European Parliament (2019)
20. ÓhÉigeartaigh, S.S., Whittlestone, J., Liu, Y., Zeng, Y., Liu, Z.: Overcoming barriers to cross-cultural cooperation in AI ethics and governance. Philos. Technol. **33**, 571–593 (2020)
21. Floridi, L.: Soft ethics, the governance of the digital and the general data protection regulation. Philos. Trans. R. Soc. A: Math. Phys. Eng. Sci. **376**(2133), 20180081 (2018)
22. Perrault, R., et al.: The AI Index 2019 Annual Report. AI Index Steering Committee, Human-Centred AI Institute, Stanford University, Stanford (2019)
23. Hawking, S., Russell, S., Tegmark, M., Wilczek, F.: Stephen Hawking: 'Transcendence looks at the implications of artificial intelligence—But are we taking AI seriously enough?' The Independent (2014). Retrieved from https://www.independent.co.uk/news/science/ste phen-hawking-transcendence-looks-at-theimplications-of-artificial-intelligence-but-are-we-taking-9313474.html
24. Polonetsky, J., Tene, O.: Privacy in the age of big data: a time for big decisions. Stanford Law Rev. Online **64**, 63–69 (2012)

25. Bessis, N., Dobre, C. (eds.): Big data and internet of things: A roadmap for smart environments. SCI, vol. 546. Springer, Cham (2014). https://doi.org/10.1007/978-3-319-05029-4
26. Oprensik, D., Taisch, M.: The value of big data in servitization. Int. J. Prod. Econ. **165**, 174–184 (2015)
27. Opresnik, D., Hirsch, M., Zanetti, C., Taisch, M.: Information – The Hidden Value of Servitization. In: Prabhu, V., Taisch, M., Kiritsis, D. (eds.) APMS 2013. IAICT, vol. 415, pp. 49–56. Springer, Heidelberg (2013). https://doi.org/10.1007/978-3-642-41263-9_7
28. Lee, K.F.: The real threat of artificial intelligence. The New York Times, 24 (2017)
29. Bostrom, N., Yudkowsky, E.: The Ethics of Artificial Intelligence. In: Frankish, K., Ramsey, W.M. (eds.) The Cambridge Handbook of Artificial Intelligence, pp. 316–334. Cambridge University Press, Cambridge, England (2014)
30. Lipton, Z.C., Steinhardt, J.: Troubling trends in machine learning scholarship. Paper presented at ICML 2018: The debates, Stockholm, Sweden (2018)
31. Acharya, A.: Why international ethics will survive the crisis of the liberal international order. SAIS Rev. Int. Aff. **39**, 5 (2019)
32. Hagerty, A., Rubinov, I.: Global AI ethics: A review of the social impacts and ethical implications of artificial intelligence (2019). https://arxiv.org/pdf/1907.07892
33. Pasquale, F.: The black box society: The secret algorithms that control money and information. Cambridge, MA: Harvard University Press (2016)
34. Burrell, J.: How the machine 'thinks': Understanding opacity in machine learning algorithms. Big Data Soc. **3**(1), 1–2 (2016)
35. Kroll, J.A.: The fallacy of inscrutability. Philos. Trans. R. Soc. A: Math. Phys. Eng. Sci. **376**(2133), 20180084 (2018)
36. Marda, V.: Artificial intelligence policy in India: A framework for engaging the limits of data-driven decision-making. Philos. Trans. R. Soc. A: Math. Phys. Eng. Sci. **376**(2133), 20180087 (2018)
37. Whittlestone, J., Nyrup, R., Alexandrova, A., Dihal, K., Cave, S.: Ethical and Societal Implications of Algorithms, Data, and Artificial Intelligence: A Roadmap for Research. Nuffield Foundation, London (2019)
38. Russell, B.: History of western philosophy and its connection with political and social circumstances. Allen and Unwin (1946)
39. Kromer, S.: Emotional intelligence: Korean nunchi across borders and identities in multinational corporations and the private sphere in the twenty-first century (Doctoral dissertation, Royal Holloway, University of London) (2017)
40. Johnston, A.I., Shen, M. (eds.): Perception and Misperception in American and Chinese Views of the Other. Carnegie Endowment for International Peace, Washington, DC (2015)
41. Capurro, R.: Privacy. an intercultural perspective. Ethics Inf. Technol. **7**, 37–47 (2005)
42. Nemitz, P.: Constitutional democracy and technology in the age of artificial intelligence. Philos. Trans. R. Soc. A: Math. Phys. Eng. Sci. **376**(2133), 20180089 (2018)

Building a Responsible AI Ecosystem: Thailand's Journey Towards Ethical AI

Apivadee Piyatumrong(✉) ⓘ

Artificial Intelligence Research Group (AINRG), National Electronics and Computer
Technology Center (NECTEC), Pathumthani 12120, Thailand
apivadee.piy@nectec.or.th

Keywords: Trustworthy AI · AI Ethics · Responsible AI · AI Governance
Landscape of Thailand

AI Thailand is a national program aiming to prepare essential infrastructure for artificial intelligence (AI) development in Thailand to promote economic growth and increase the country's competitiveness [1]. Multiple facets of infrastructure, human capacity, and a practical ecosystem need to be established. As a result, Thailand's national AI strategy and action plan (2022–2027), known as NAIS, was approved by the Prime Minister's Cabinet Office on July 26, 2022. The strategy aims to provide Thailand with an effective ecosystem for AI development. Accordingly, by implementing its strategy, NAIS will concentrate on enhancing the economy and improving Thai people's quality of life by 2027. There are five strategies and fifteen work plans, and this work pays attention to the first strategy that prepares Thailand's readiness in social, ethical, law, and regulation for AI applications.

Before we explore the AI ethical landscape, it is crucial to grasp the unique characteristics of generative AI. Unlike other AI applications, generative AI is a distinct subset of artificial intelligence that has the ability to create new content based on patterns it has learned from existing data, such as text, images, audio, and video. This innovative AI is made possible through advanced machine learning techniques, including Generative Adversarial Networks (GANs) and transformer-based models. These models are trained on extensive datasets to identify and replicate patterns, enabling them to generate original and human-like content. This unique capability sets generative AI apart from other types of AI, which primarily focus on analyzing or predicting based on existing data.

Generative AI has significantly impacted daily life by enhancing creativity and productivity across various domains. For instance, it can personalize entertainment by curating music and movies, assist in meal planning by suggesting recipes based on available ingredients, and improve customer service through AI-driven chatbots. Additionally, it aids education by providing instant answers and homework help and even influences personal relationships through AI-generated virtual companions. However, generative AI raises ethical concerns about creating fake content, deepfakes, misinformation, and more. Serious misinformation by generative AI in this early technology adoption stage

© IFIP International Federation for Information Processing 2024
Published by Springer Nature Switzerland AG 2024
R. M. Davison and D. Kreps (Eds.): HCC 2024, IFIP AICT 719, pp. 9–11, 2024.
https://doi.org/10.1007/978-3-031-67535-5_2

significantly impacts human value on many occasions [3–5]. AIAAIC has collected AI-related incidents since 2012 [6], and the latest HAI 2023 report [7] highlights an increase in incidents since 2017 that reached 260 incidents in 2021, underscoring the potential risks associated with AI. Furthermore, many guidelines and approaches to governing AI have been published, with hundreds of documents addressing ethical principles and best practices since 2016 [8–10] confirming the concerns about AI risk. Accordingly, over 123 bills related to AI have been introduced globally [7], reflecting the growing legislative focus on AI governance.

To be prepared for privacy, safety, security, accountability, transparency, explainability, fairness, and non-discrimination issues from adopting AI technology, the nation must prepare itself by utilizing different means such as laws, regulations, guidelines, or recommendations. Ministry of Digital Economy (MDES) launched the AI Ethics Guideline [11] and was approved by the cabinet to be adopted widely in Thailand since February 2021. The Electronic Transactions Development Agency (ETDA) published the Thailand AI governance guideline [2] for organizations to study and adapt to its suitability. Furthermore, draft bills were proposed and opened to the public for study. These include (1) Draft Royal Decree on Business Operations that uses Artificial Intelligence Systems proposed by the Office of the National Digital Economy and Society Commission (ONDE) & Chulalongkorn University, (2) Draft Act on Promotion and Support for National AI Innovations proposed by ETDA & Thammasart University, (3) Draft Notification on AI Sandbox 2023 proposed by ETDA, (4) Draft Notification on Criteria and Method for Assessing and Managing Risks Arising from the Use of AI 2023 proposed by ETDA. Apart from that, several collaborations are being conducted to deliver AI standards for local businesses to gain customers' trust. A memorandum of understanding among sixteen organizations, both private and public organizations, was signed on developing Thai AI standards, which are currently in development [12]. Moreover, several ethical guidelines for the private sector and organizations have been proposed, such as the NSTDA AI Ethics Guideline (2022) by the National Science and Technology Development Agency (NSTDA), the Regulatory Framework for the Use of AI/ML in the Capital Market (2023) by the Securities and Exchange Commission (SEC), and the Thailand Artificial Intelligence Guidelines 1.0 (2022) by Chulalongkorn University.

In summary, Thailand is working to govern AI through regulatory guidelines and cabinet commitments. However, specific AI laws have yet to be studied carefully and processed openly and inclusively, particularly in unique sectors such as medicine, finance, and investment. The legislative process for those draft bills is currently in progress, with studies being conducted on various aspects to develop comprehensive AI regulations.

References

1. AI Thailand by the Assistant Secretary of Thailand National AI Strategy steering committee. https://ai.in.th. Accessed 17 May 2024
2. Thailand AI Governance Guideline for Executives. https://www.etda.or.th/th/Our-Service/AIGC/index.aspx. Accessed 17 May 2024
3. How Wrongful Arrests Based on AI Derailed 3 Men's Lives. https://www.wired.com/story/wrongful-arrests-ai-derailed-3-mens-lives/. Accessed 05 Feb 2024

4. Child sexual abuse content growing online with AI-made images. https://www.theguardian.com/technology/2024/apr/16/child-sexual-abusecontent-online-ai. Accessed 18 May 2024

5. Cruise recalls all self-driving cars after grisly accident and California ban. https://www.theguardian.com/technology/2023/nov/08/cruise-recall-self-drivingcars-gm. Accessed 18 May 2024

6. AIAAIC is an independent, non-partisan, grassroots public interest initiative thatexamines and makes the case for real AI, algorithmic, and automation transparency and openness. https://www.aiaaic.org/home. Accessed 18 May 2024

7. 2023 State of AI in 14 Charts - A snapshot of what happened this past year in AI research, education, policy, hiring, and more., https://hai.stanford.edu/news/2023state-ai-14-charts. Accessed 18 May 2024

8. Principled Artificial Intelligence – Mapping Consensus in Ethical and Rights-Based Approaches to Principles for AI. https://cyber.harvard.edu/publication/2020/principled-ai. Accessed 18 May 2024

9. Jobin, A., Ienca, M., Vayena, E.: The global landscape of AI ethics guidelines. Nat. Mach. Intell. **1**, 389–399 (2019)

10. Soudi, M., Bauters, M.: AI guidelines and ethical readiness inside SMEs: a review and recommendations. DISO **3**, 3 (2024)

11. Thailand AI Ethics Guideline. https://drive.google.com/drive/mobile/folders/1kB7dxS5nXibeX3UwP8ABQMlrO44xm41f. Accessed 18 May 2024

12. AI Thailand Forum 2023 Embracing the Future of AI. https://www.aithailandforum.com/agenda. Accessed 18 May 2024

Historical Characteristics of Artificial Intelligence and Robotics Research in Japan in Relation to Postwar Political and Cultural Background

Mai Sugimoto(✉) (iD)

Kansai University, Osaka, Japan
`msgmt@kansai-u.ac.jp`

Abstract. Artificial intelligence (AI) and robotics research in Japan do not necessarily follow the same trends as those in the United States of America. This is because military research in Japan has been severely restricted since the end of World War II, and the research funds available for defense expenditures are limited. The development of AI in Japan during the second AI boom, including the Fifth Generation Computer System, was intended for industrial promotion and international cooperation. Additionally, Japanese researchers in AI and robotics have been positively influenced by the popular subcultures of manga and anime. This has led to a trend in the history of Japanese robotics toward developing human-like bipedal robots and robots that can spend time together at home. This study examines the position of military research in Japan and analyzes the characteristics of AI research during the second AI boom, which marked the beginning of AI research in the country. It also explores how the political and cultural context in Japan influenced Japanese researchers and engineers.

Keywords: Artificial Intelligence (AI) history · military research · Fifth Generation Computer System (FGCS) · Japan

1 Introduction

While artificial intelligence (AI) research is expanding globally, the focus of AI and robotics research in Japan has been somewhat different from that of other countries. In the US, AI research, which is historically closely related to computer development, cybernetics research, and computer science, has been largely funded by Defense Advanced Research Projects Agency (DARPA), and the military aspect of AI has had to be considered significant[1][1–3]. However, in Japan, public criticized research involving military, and the amount of defense expenditure and large-scale funding for research not directly

---

[1] The development of computing machines and analog feedback devices in the mid-20th century had a military background [1]. Edwards discussed the connection between AI research and military technology development, backed by DARPA funding during the Cold War [3].

© IFIP International Federation for Information Processing 2024
Published by Springer Nature Switzerland AG 2024
R. M. Davison and D. Kreps (Eds.): HCC 2024, IFIP AICT 719, pp. 12–22, 2024.
https://doi.org/10.1007/978-3-031-67535-5_3

related to industry was limited [4]. As a result, Japan has not followed this global trend completely, but develop with a certain degree of originality.

The history of AI has seen three major booms: 1. The success soon after the Dartmouth conference in the late 1950s and the 1960s, 2. The period of expert systems in the late 1970s and the 1980s, and 3. The rise of machine learning from the mid-2010s [6]. They were started in the US and fueled by significant investments from both the government and business sectors, which expected the industrial and military applications of AI to be useful. In Japan, while investments were made in the industrial sector, there was no research funding with defense implications, and virtually no military/defense-related research was conducted.

AI researchers in Japan, at present, do not view the history of AI research in Japan as tied to the military and defense. For instance, *the Journal of the Japanese Society for Artificial Intelligence* (JSAI) published a cover series in 2023 that depicts an illustrated scroll of AI history, known as the AI history of *emaki* in Japanese, wherein six covers are connected in sequence. *Emaki* is a traditional Japanese-style painting that presents a story and its timeline on a long horizontal roll. The 2023 JSAI journals depict historical AI events from the 1940s to the present day. The first issue's cover depicts the first boom in the 1950s and the 1960s, primarily centered in the US, such as the Dartmouth conference [6]. At the time, AI research had not yet begun in Japan. Starting from the second issue, which portrays the second boom in the 1980s, when AI research was initiated in Japan, there are more illustrations of AI research in Japan. Elements related to the military, such as DARPA and autonomous weapons, have been reduced, focusing on robots and games in Japan [7] (Fig. 1). In the sixth issue, the future vision is presented, which still lacks the depiction of military-related elements [8]. During the discussion of the cover art project's supervision team, which mostly included AI researchers, it was noted that few elements depicted in the cover art were related to military and defense [9]. The Ethics Committee at the JSAI national conventions has been discussing the relationship between AI and the military and defense as a global issue since the late 2010s [10]. However, this does not appear to be a pressing issue in Japan.

This is probably due to the historical background of Japan's political and cultural position after World War II. This study examines the position of military research in Japan and analyzes AI research characteristics during the AI boom in the 1980s, which marked the onset of AI research in the country. It also investigates the ways in which Japan's political and cultural context influenced its researchers and engineers.

2 Prohibition of Military Research in Japan in the Latter Half of the 20th Century

Post World War II, the Allied Powers occupied Japan, and the General Headquarters, the Supreme Commander for the Allied Powers (GHQ/SCAP) banned military research in the country, including research on atomic energy and aircraft development, until the Treaty of Peace with Japan became effective in 1952. The new Constitution of Japan, enforced in 1947, renounced war in Article 9. Until the early 1950s, Japan did not have any military organizations, not even the Self-Defense Forces. In 1952, research restrictions were lifted. However, the Japanese academic community strongly criticized any

Fig. 1. Cover pages of Journal of Japanese Society for Artificial Intelligence in 2023 (No.1 and No.2)

research associated with the military. The Science Council of Japan was established by the academic community in 1949. The council issued "Statement of Determination to Never Become Engaged in Scientific Research for War Purposes" in 1950 and "Statement of No Scientific Research for Military Purposes" in 1967. Consequently, Japanese academia consistently refrained from or prohibited involvement in military research post World War II, even after postwar reconstruction [11]. Until the 21st century, the Self-Defense Forces maintained limited cooperation with the defense industry. Japan allocated only 1 percent of its gross domestic product to the military, resulting in a restrained budget.

Japanese society also did not allow research related to the military. Japanese Government's defense research budget, as well as the US military's funding of Japanese academia, was a subject of criticism from Japanese society, the mass media, and the academic community. For example, newspapers reported that the executive committee of an international conference on semiconductors held in Kyoto in 1967 received financial support from the US Army Research Office Far East, and it was denounced by the budget committee of the Japanese diet. Academic societies strongly criticized the fact that the Physical Society of Japan was not informed of this. In the same year, the Physical Society of Japan announced that it would "never accept any assistance or any other cooperative relationships from the military, whether domestic or foreign" [12].

Thus, in the 20th century after the World War II, the Japanese government did not allocate large amounts of money directly to the military. Instead, they assigned funds to promote domestic industries. This strategy boosted Japan's economic growth in the 20th century. Civilian companies were responsible for developing and manufacturing military technology products, including semiconductors, for use by the Self-Defense Forces or for sale to the US military. In Japan, large grants by the government were usually for industrial applications, with relatively little financial support for academic research with few implications for domestic industry. This directed academic researchers towards research that was possible with little funding or based on their own cultural interests. These factors have influenced the research and development of AI and robotics in Japan.

3 The Fifth Generation Computer Systems Project and the Second AI Boom

AI research in Japan began in the late 1970s, before the second AI boom. In 1977, young Japanese researchers organized an AI reading group called AIUEO inspired by an AI seminar at Essex University. They read various AI-related papers and books, including works by Minsky and Winograd, and those published in the International Joint Conference of Artificial Intelligence (IJCAI) [13]. In the 1970s and the 1980s, no Japanese university systematically taught AI. Therefore, it can be said that AI research in Japan began as academic research, and AI research was not even institutionalized at that time, leaving no room for military implications.

The Fifth Generation Computer System (FGCS) project is often considered an early development of AI in Japan. However, examining Japan's approach to industrial development is necessary to assess FGCS' historical significance. Post World War II, the Japanese government provided substantial subsidies to manufacturing industries, including the automobile and heavy chemical industries. From the late 1960s, the Japanese government strongly encouraged the development of the semiconductor and computer industries, promoting more than 10 new technology development projects until the 1990s [14]. The Ministry of International Trade and Industry (MITI) funded several companies to develop Japanese computers and semiconductors. Until 1974, the Japanese government limited the number of integrated circuit chips that US companies could sell in Japan. Even after this limit was removed, Japanese companies did not purchase semiconductor chips from the US [15]. Consequently, Japan gained a large share of the global semiconductor market. MITI actively encouraged cooperation among semiconductor manufacturers, leading to the establishment of the VLSI Technology Research Association in 1976. In Japan, the budget was not allocated to military spending, but was devoted to boosting economic growth. This approach was successful.

The FGCS project was MITI's initiative to foster the computer industry. In the late 1970s, MITI was searching for new computing systems. The FGCS project was based on the idea of creating a computer with a new architecture that could effectively introduce very-large scale integration (VLSI) technology, respond to software crises, and handle nonnumeric data effectively [16]. Since the industrial espionage case against IBM in 1982, Japan could no longer cut into its market share with IBM-compatible computers and needed a new domestic Japanese computer [17]. The FGCS was not initially an AI project but rather a project created to foster the semiconductor and computer industries in Japan. At the time of its inception, the wave of AI as expert systems had not yet arrived in Japan. Although divergent from its current general interpretation, the FGCS project members repeatedly told at the time [18–20], and recalled in later years [21], that the FGCS project was not a project for AI.

The FGCS project was led by the Institute for New Generation Computer Technology (ICOT), founded in 1982. The ICOT was organized by researchers under the age of 35 from 8 computer manufacturers (Fujitsu, Hitachi, NEC, Toshiba, Mitsubishi, Oki, Matsushita, and Sharp), the Nippon Telegraph and Telephone Public Corporation, and the National Institute of Advanced Industrial Science and Technology [16, p. 126]. The FGCS project lasted 11 years, from 1982 to 1992, and received approximately 60 billion JPY in government funding. Combined with corporate investments, approximately 100 billion JPY was spent [17, pp. 157–158]. The second AI boom, which began in earnest around 1984, also helped this project. Finally, a prototype system with a parallel inference machine and the parallel logic programming language KL1 at its core was developed, making it one of the fastest inference machines in the world [22].

To some extent, the FGCS was an internationally open project. In 1981, a year before the launch of the project, an international conference was held, and MITI invited the governments of Europe and the US to participate, resulting in over 80 participants from various countries. Toru Moto-oka, a professor at University of Tokyo and one of the leaders of the FGCS project, recalls, "In the case of the US, the Department of Defense funds projects, which means that people working on such projects might hesitate to share information with other countries. If it were happened, it could negatively affect international cooperative development. As this is fundamental research, it is hoped that a worldwide system can be established where countries share their research results with each other for the cooperative development of computers that will contribute to humanity in the future" [16, p. 14]. Moto-oka also suggested that the realization of machine translation could eliminate the language barriers in Japanese and make international contributions easier. This was one of the most important themes in Japanese AI research, especially in Japan, where there were few English speakers. Thus, one of the social roles of the FGCS project was to contribute to the international community by developing the information industry [16, p. 27]. In other words, its purpose was not only to ensure Japan's economic competitiveness on the international stage but also to indirectly promote Japan's security facilitated by international exchange through information technology projects. This contrasts with the intent of a typical military technology project.

Meanwhile, in 1983, DARPA initiated the Strategic Computing Initiative in the US to compete with the FGCS project. The Microelectronics and Computer Consortium (MCC) also worked on Cyc. Bobby Ray Inman, a Navy General and former National Security Agency director, served as the head of the MCC in 1983. It is clear that these organizations were assembled in connection with the Department of Defense and military. The Alvey Project (1983–1987) was launched in the United Kingdom to develop a new computer that included a knowledge base, software engineering, and user interfaces. Additionally, the European Strategic Programme on Research in Information Technology was launched in Europe [16, pp. 9 and 11; 23].

Feigenbaum and McCorduck's *The Fifth Generation* (1983) highlights the view of the US regarding AI and semiconductors at the time, which was different from the one of Japan [23]. The book portrays the FGCS as a knowledge information processing system and a potential threat to the dominance of the US computer industry. The authors express skepticism toward Japan's positive attitude toward international cooperation, stating that it may be a "certain amount of pious lip service" [23, p. 133]. In Chapter 6, they also discuss military national security issues arising from the FGCS project. This was because of Japan's history of exporting semiconductors to the US since the 1970s, which led to an increased reliance on foreign-made semiconductors in the US military, a problem that had been identified [15]. In the late 1970s and 1980s, Japan and the US had a tense relationship over semiconductors. This was considered an economic and security conflict in the US. Seemingly, the US viewed Japan's new computer systems as a security concern.

However, FGCS-related articles and books published in Japan have little discussion on military or national security issues. Since the end of the war, the Japanese government had largely limited its defense expenditures, focusing instead on economic growth and promoting government-led intercompany cooperation. The FGCS project is an example of this. Researchers involved in the FGCS project emphasized international cooperation. The international conference on FGCS was held, and Japanese, European, and US researchers engaged in active discussions. The ICOT team recognized the need for international contributions, and the FGCS project was envisioned to address issues such as an aging society, which became important for subsequent AI and robotics in Japan. Thus, even if one considers the FGCS project to be an early development of AI in Japan, it was detached from the military aspects of AI development and had a research orientation supported by the Japanese cultural and social context.

4 AI, Robotics, and Researcher Community in Japan During 1980s–2000s

Academic organizations related to AI and robotics were not well-developed in Japan until the 1980s. One AI researcher recalled that senior computer researchers at the time held negative perceptions toward AI research [13]. In 1983, the Robotics Society of Japan and the Japanese Cognitive Science Society were established, followed by the establishment of the JSAI in 1986. The AI researchers active since the 1980s had nurtured their interests through informal reading circles since the 1970s. This was how AI research began in Japan.

Robots were already being used in the industry in the 1980s, and Japan had a large market for industrial robots, with shipments amounting to JPY 600 billion per year by 1991 [24]. But in terms of non-industrial applications, a robotics researcher Toshio Fukuda, the first Asian president of the IEEE, noted that Japanese robotics research in the 1980s differed from trends in the US, and publications were not readily accepted [25]. This was because of insufficient funding for robotics research in Japan, making it impossible to acquire expensive equipment such as DEC minicomputers, programmable universal manipulator for assembly (PUMA) robots, and vision cameras, which were standard in the US. Japanese researchers had to design their own manipulators and mechanisms, develop new control methods using 8/16-bit PCs, and conduct unique studies. Robotics researchers in the US often received funding from the DARPA [26]. However, in Japan, the AI and robotics research with no industrial application received no funding. Thus, Japanese academic researchers conducted studies to achieve their own ideal AI and robotics without military direction.

Japanese researchers in the fields of AI and robotics have also been positively influenced by the popular subcultures of manga and anime. Specifically, Osamu Tezuka's manga *Astro Boy*, which features a cute, human-like robot boy who can coexist with humans, significantly impacted senior Japanese researchers. Traditionally, the Japanese public has viewed robots, such as industrial robots, as devices that just perform exactly what they are supposed to do. It was not until the late 20th century that the public began to view robots as potential human partners [27]. However, by the 1980s, researchers involved in the development of robots considered friendly, at-home, human-like communicative robots depicted in manga and anime, such as Astro Boy, as desirable goals for robots. In 1985, for example, an article introducing robotics technology in *AI Journal* published in Japan described robots as "eventually entering the home" [28]. Also, Tomomasa Sato, who later became a robotics researcher at the University of Tokyo, stated in 1987 that "it is desirable for robots to leave the factory and eventually be able to work in the home" [28]. This could be called a "cultural imperative"[2] as discussed by Schiffer [30].

During a conversation with Tezuka in the 1980s, robotics researchers made statements regarding Astro Boy and their work [31]. Yuji Hosoda, who worked for Hitachi, stated that "I develop robots with the belief that the final form of a robot is still Astro Boy." Similarly, Masaru Uchiyama of Tohoku University suggested that development efforts were driven by fictional images, such as those of the Astro Boy, in addition to economic principles. Astro Boy was featured in a JSAI special issue in 2003, where it was noted that this manga contributed to the development of robot technology [32]. Hitoshi Matsubara, one of the founders of RoboCup, has stated in papers published in JSAI journals that he has been a fan of Astro Boy since childhood, and that his goal is to create a robot like Astro Boy [33, 34].

In addition to Astro Boy, Japanese manga has a lineage of friendly robots such as Doraemon (serialized from 1969–1997), for young children, which has a character recognition rate of 97% in a national survey in Japan. These works potentially influenced the younger generation of the Japanese public [35]. There are many other popular works related to war and robotics/AI in Japan, such as the *Gundam* series, the *Macross* series, and *Ghost in the Shell*. However, most of these focused on older audiences, including teenagers. In contrast, the main characters in robot manga and anime, popular across all age groups, are friendly robot partners. This has boosted the image of robots among the younger generation in Japan and influenced robotics research in the country. In 2016, a survey conducted by the Ministry of Internal Affairs and Communications found that in Japan, more people perceive AI as "technology that gives computers an ego (emotion)" [36] as compared to the US.

In Japan, this has led to the development of human-like bipedal robots and robots that can spend time together at home. HONDA's biped humanoid P2 in 1997 and Sony's AIBO are early examples of this trend. In the 1990s, some scholars considered bipedal

---

[2] According to Schiffer, the cultural imperative "is a product fervently believed by a group -- its constituency -- to be desirable and inevitable, merely awaiting technological means for its realization." [28]

humanoids as best suited for providing services in the tertiary industry [37]. Documents of the time also state that these bipedal robots were created to learn human movements and to be used in daily life. In Japan, walking robots were not intended for military use but rather to be integrated into daily life, to assist in caring for older adults, or contribute to disaster relief. Japan hosts various robotic contests, such as the RoboCup and rescue robot contests, which are often broadcast on TV. This may be due to the favorable perception of robotics and its recognized significance as an industry and in society. The perception of robotics among Japanese researchers in the 20th century and the perception fostered among the general public from an early age coincided with and influenced research trends in Japan. It could be said Japan was a niche with limited military implication and this have given rise to a unique cultural imperative.

5   Change in the 21st Century?

Until the 20th century, research on AI and robotics in Japan, particularly in areas related to the semiconductor and computer industries, was conducted through collaborative projects among several companies led by MITI, such as the FGCS project. These projects were mainly aimed at industrial promotion. As previously mentioned, academic research conducted at universities was distant from the military.

However, in the 21st century, the Abe administration implemented a policy to increase the budget for the Self-Defense Forces and involved academia in the development of military technologies. The first step in this direction was performed in the field of space research. Since the 1960s, space research and development for peaceful purposes had been firmly established, but in 2008, the Basic Act on Space abolished this principle. Since 2012, the Japanese Aerospace Exploration Agency, together with the Ministry of Defense has been responsible for Japan's national security. In 2014, the embargo on weapons was lifted, and the government promoted the use of civilian technologies that could be applied to defense technologies. The National Security Technology Research Promotion Fund was established in 2015. Since then, defense budgets have been allocated to academic research, including AI and robotics [38]. The budget for 2017 amounted to 11 billion JPY.

Since the late 20th century, the government budget allocations to universities have declined, resulting in a severe shortage of funding for university researchers. Consequently, researchers have turned to US military grants, which have been awarded to Japanese academic researchers since the late 20th century. Since 2008, researchers in the fields of lasers, sensors, and machine learning have been granted 880 million JPY for 9 years [38]. From 2010 to 2016, 11 of these researchers, including those working in AI and lasers, received grants from the US Air Force and Navy totaling 200 million JPY, or 1.5–45 million JPY per researcher [39]. These researchers received grants because the government provided limited funding. They were informed that the results would not have any direct military application [38].

Criticism of the relationship between academic research and military in domestic academia has been intense. Since the Science Council of Japan released its "Statement on Research for Military Security" in March 2017, security-related research has not been actively promoted in the academic community. Furthermore, owing to the absence of military research in Japanese academia for many years, Japanese researchers have had limited opportunities, resulting in a lack of concrete discussions on military applications of technology. The Japanese academic researchers have applied for funding from the National Security Technology Research Promotion Fund and the US military since the 21st century because of the annual reduction in government funding for pure scientific research.

Although Japan's defense budget and discussions about AI and the military have increased since the beginning of the 2010s, Japanese researchers do not seem to believe that they can be actively involved in the development of such technology. But changes in public demands might affect research trends. The global propagation of AI and information technology is undeniable; however, considering the impact of local political and cultural circumstances on research and development is crucial.

Acknowledgments. I thank the anonymous reviewers for the careful reading of my manuscript and their insightful comments and suggestions.

References

1. Mindell, D.A.: Between Human and Machine: Feedback, Control, and Computing before Cybernetics. Johns Hopkins University Press, Baltimore (2002)
2. Roland, A., Shiman, P.: Strategic Computing: Darpa and the Quest for Machine Intelligence, 1983–1993. The MIT Press, Cambridge, Massachusetts (2002)
3. Edwards, P.N.: The Closed World: Computers and the Politics of Discourse in Cold War America. The MIT Press, Cambridge, Massachusetts (1996)
4. Interview with Koji Yada: AI Journal **3**, 54–61 (1986). (in Japanese)
5. Sugimoto, M.: What is the "AI booms"?. Gendai Shisou 現代思想, April 2019, 64–68 (2019) (in Japanese)
6. Cover page. Journal of the Japanese Society for Artificial Intelligence 38(1) (2023)
7. Cover page, Journal of the Japanese Society for Artificial Intelligence 38(2) (2023)
8. Cover page, Journal of the Japanese Society for Artificial Intelligence 38(6) (2023)
9. Sakaki, T., et al.: Cover project round-table discussion: the completion of the scroll of AI history. J. Jpn. Soc. Artif. Intell. **38**(6), 980–989 (2023) (in Japanese)
10. JSAI website. Ethics Committee Archives. https://www.ai-gakkai.or.jp/ai-elsi/. Accessed 26 Jan 2024
11. Sugiyama, S.: "Gunji Kenkyu" no Sengo Shi 「軍事研究」の戦後史 [Postwar Debate on Military Research], Minervashobo, Kyoto (2017). (in Japanese)
12. Konuma, M.: The Science Council of Japan in the Early Days on Military Research. Gakujutsu no Doukou 学術の動向 [Trends in the Sciences], **22**(7), 10–17 (2017). (in Japanese)
13. Saito, Y., et al.: AIUEO: from the beginning to the end. J. Jpn. Soc. Artif. Intell. **35**(2), 257–260 (2020). (in Japanese)
14. Takaishi, Y.: Development of the computer industry and industrial policy in Japan. Ann. Soc. Ind. Stud. Jpn. **3**, 24–38 (1988). (in Japanese)
15. Miller, C.: Chip War: The Fight for the World's Most Critical Technology. Scribner, New York (2022)
16. Moto-oka, T., Kitsuregawa, M.: Dai Go Sedai Computer 第五世代コンピュータ (The Fifth Generation Computer). Iwanami-shoten, Tokyo (1984). (in Japanese)

17. Hasida, K.: Twenty-year afterthoughts on ICOT (<Special Issue>the fifth generation computer systems project and the future of AI). J. Jpn. Soc. Artif. Intell. **29**(2), 157–158 (2014). (in Japanese)

18. Yokoi, T.: AI Kakuron no Susume AI各論のすすめ [Itemized Discussion for AI]. AI J. **8**, 6–8 (1987). (in Japanese)

19. Shimada, S.: 1987 Nen no Hatsuyume aruiha 2 Nen Go no Yosoku 1987年の初夢あるいは2年後の予測 [First Dream of 1987 or Prediction for the Next Two Years]. AI J. **8**, 8–11 (1987). (in Japanese)

20. Yokoi, T., et al.: Zadankai Nihon Gata Project Ron Producer Ron 座談会 日本型プロジェクト論、プロデューサー論 [Round-table Japanese Project Theory and Producer Theory]. AI J. **12**, 20–31 (1987). (in Japanese)

21. Hayashi, S.: The Fifth Generation Computer Systems Project: Japan in the 1980s. Rekihaku, Special Feature: A Contemporary History of Artificial Intelligence, pp. 39–46 (2022). (in Japanese)

22. Yamaguchi, T.: What Should We Do for Future AI Projects Based on FGCS Reviews? (<Special Issue>The Fifth Generation Computer Systems Project and the Future of AI). Journal of the Japanese Society for Artificial Intelligence **29**(2), 115–119 (2014). (in Japanese)

23. Feigenbaum, E.A., McCorduck, P.: The Fifth Generation: Artificial Intelligence and Japan's Computer Challenge to the World. Addison-Wesley Publishing Company, Reading, Massachusetts (1983)

24. Asama H.: Robotics in Heisei : a review: trends and future prospects of robot technology. J. Jpn. Soc. Precis. Eng. **86**(1), 23–27 (2020). (in Japanese)

25. Fukuda, T.: Robotics and my research inspiration. J. Robot. Soc. Jpn. **24**(3), 333–337 (2006). (in Japanese)

26. Singer, P.: Wired for War: The Robotics Revolution and Conflict in the Twenty-fist Century. Penguin Press (2009)

27. Sena, H.: Robot dreams, robot visions: a social view to robotics. J. Robot. Soc. Jpn. **20**(6), 610–614 (2002). (in Japanese)

28. Technology, R.: AI Journal **1**, 50–52 (1985). (in Japanese)

29. Q&A on AI: AI Spectrum **2**, 15–46 (1987). (in Japanese)

30. Schiffer, M.B.: Cultural imperatives and product development: the case of the shirt-pocket radio. Technol. Cult. **34**(1), 98–113 (1993)

31. Tetsuwan Atom no Sekai to Robot Gijutsu 鉄腕アトムの世界とロボット技術 [The World of Astro Boy and Robotics]. J. Robot. Soc. Jpn. **4**(3), 306–311 (1986). (in Japanese)

32. Sugawara, S.: Life-Style and Mind of Astro-Boy (<Special Issue>Astro-Boy). J. Jpn. Soc. Artif. Intell. **18**(2), 153–158 (2003). (in Japanese)

33. Matsubara, H.: Editor's Introduction to "Astro-Boy" (<Special Issue>Astro-Boy). J. Jpn. Soc. Artif. Intell. **18**(2), 144 (2003). (in Japanese)

34. Matsubara, H.: Astor-Boy from a Viewpoint of Artificial Intelligence (<Special Issue>Astro-Boy). J. Jpn. Soc. Artif. Intell. **18**(2), 159–162 (2003). (in Japanese)

35. Kanoh, H.: Elementary school students' awareness of AI and robots. Res. Rep. Inf. Educ. **2**, 9–16 (2020). (in Japanese)

36. Ministry of Internal Affairs and Communications, Japan: Figure 4–2–1–3. 2016 White Paper on Information and Communications in Japan. https://www.soumu.go.jp/johotsusintokei/whitepaper/h28.html. Accessed 27 Jan 2024

37. Japan Science and Technology Agency: Madamada Korekara: Hitogata Robot ni Kakeru Yume まだまだこれから:人型ロボットにかける夢 [Way to Go: Dream for Humanoid Robot]. JSTnews **2**(7), 4–7 (2005). (in Japanese)

38. Kabata, H.: Military application of artificial intelligence. J. Sci. Technol. Stud. **16**, 31–42 (2018). (in Japanese)
39. Saito, Y.: Beigun Shikin Mondai to Nichibei Nikoku Kan Gijutsu Kyoryoku ni Kakawaru Ugoki 米軍資金問題と日米二国間技術協力に関わる動き [US Funding Issue and Developments related to US-Japan Bilateral Technical Cooperation]. J. Jpn. Sci. **52**(11), 47–39 (2017). (in Japanese)

The History of AI in Thailand: Thickening Our Vision of AI by Caring for a Marginalised Actor

Soraj Hongladarom(✉) ⓘ, Auriane van der Vaeren ⓘ, and Suppanat Sakprasert ⓘ

Center for Science, Technology, and Society, Chulalongkorn University, Bangkok, Thailand
Soraj.H@chula.ac.th, sderoid.xi@gmail.com

Abstract. The first steps of artificial intelligence (AI) in Thailand can be traced back to Thai university course material from 1975. Around the mid-1980s, the Thai government started to acknowledge the potential of AI technology to become a driver for economic development. Yet, despite the benefits of monitoring AI's domestic developments, such research remains scarce for Thailand. This paper seeks to fill this current literature gap through offering a centralised compilation of information about the unfolding of AI technology in Thailand from its beginnings to this day. Given the scarcity in written forms of publicly available knowledge, we enrich the available literature with three qualitative expert interviews, and we contextualise Thailand's case within the broader Southeast Asian and Asian context. Thailand is regularly overlooked or marginalised when we 'think AI'. However, we will understand that Thailand's role in AI's history is critical if willing to 'thicken' existing descriptions of AI. Reductivist understandings of AI might speak of a technology transfer of the 1970s from the Occident to Thailand and stop there. However, firstly, this obscures China's growing influence over Thailand's AI development since 2015. Secondly, this obscures Thailand's necessarily participative role in AI's development—both within Thailand and abroad—which thus contrasts with the understanding where Thailand would purely be subjected to 'foreign' influence in this technology transfer. Reviewing the history of AI in Thailand allows to add critically missing layers to our currently too partial understanding of the history of AI globally. We also hope to insufflate new air into this musty research avenue to further improve our understanding of Thailand's contemporary sociotechnical unfoldment.

Keywords: Thailand · artificial intelligence (AI) · AI history · technology transfer · sociotechnology

1 Introduction

Today, artificial intelligence (AI) is veritably inundating our everyday lives. A record of the introduction of AI in Thailand can be found in university course material from 1975, when AI began to be taught at Thai universities [1]. It was however not until the late 1980s that research and innovation in AI started to become more serious, as in having Thai academics themselves innovate the technology [1, 2]. Around the mid-1980s, the

© IFIP International Federation for Information Processing 2024
Published by Springer Nature Switzerland AG 2024
R. M. Davison and D. Kreps (Eds.): HCC 2024, IFIP AICT 719, pp. 23–45, 2024.
https://doi.org/10.1007/978-3-031-67535-5_4

Thai government started to acknowledge the potential of AI technology to become a driver for economic development; creating in 1986 a research agency to stimulate, among other things, research in AI. It was further not until the mid-to-late-1990s that the private sector started to gradually assimilate AI into its activities and to innovate by developing AI systems "from scratch" [3]. Comparatively, in the Western world, AI had been developed in the 1940s and had become a standalone scholarship in the US already in 1956 [4, 5]. Today, well-aware that "AI is expected to add 15 trillion U.S. dollars to the world economy over the next decade" [6], AI has become a dedicated field of research and technological innovation in Thailand in both academia and the private sector. However, Thailand still has a long way to go if willing to become a global leader [7–10].

Seeking to fill in the void that currently characterises research on the development of AI technology in Thailand, we here engage in an in-depth literature review complemented with interviews with three Thai AI pioneers. This complementation will show to be critical as existing literature will show to be minimal. Asanee Kawtrakul and Prasong Praneetpolgrang are the first to offer a historical review of AI's development in Thailand [1]. To this day, despite that AI has engulfed the world, it remains the only such publication. In seeking to understand AI's development in Thailand, the ambition of our paper is thus to pursue this earlier move. Our ambition is thus not about seeking to make any sort of groundbreaking theoretical or analytical advance. Rather, more humbly, our ambition is to further compile and centralise information on the development of AI in Thailand, from 1975 to this day, to facilitate future research.

"[T]he introduction of some set of material artifacts out of their original context of human praxes or techniques, into some other cultural context" is what Don Ihde called a 'technology transfer' [11]. Looking at the events that characterise AI's contextual development in Thailand is about nurturing an attentiveness to the peculiarities of the domestic adoption of what was a culturally 'foreign' technology. Such attention enables us to critically 'thicken' existing descriptions of AI [12], that still today overlook the participative role of Thailand in orienting the course of the development of AI both domestically and abroad. This 'thickness' refers to María Puig de la Bellacasa's care for the way things are known and thought in different worlds [12]. Few would be the people outside Thailand who would think of Thailand upon 'thinking AI'. Yet, paying attention to this marginalised country will reveal to be critical if willing to craft a more in-depth understanding of the development of AI globally. In other words, it will reveal to be critical if willing to thicken our vision of AI with this technology's characterising layers.

2   Methodology

If one seeks to understand the historical development of AI technology in Thailand, one is quickly hit by the scarcity of such overview and quickly comes to the realisation that this quest requires far-reaching research on the Web and arduous networking efforts to pull this information together (this also reflects in the length of our reference list). Boonserm Kijsirikul and Thanaruk Theeramunkong's quantitative surveys of Thai companies, institutes, and universities in 1992 and 1999 seem to be the first available information on how AI technology was used and developed in Thailand [3]. From 1999, we seemingly needed to wait until 2014, when Kawtrakul and Praneetpolgrang offer the first compilation of AI's historical development in Thailand [1]. To this day, it remains the only such publication. The number of publicly available reports on the digital transformation of Thailand and recent AI developments are lately on the rise[1]; yet, very few exist in, say, pre-2015 times[2].

Seeking to provide an overviewing compilation of the historical development of AI technology in Thailand, we decided to 'thicken' the existing descriptions in three ways. Firstly, we enrich the available literature with three qualitative expert interviews. Secondly, adhering to the feminist premise that phenomena come into existence through their relations [12], we include key events in Thailand's history of computation and the Internet to better weave together the relations that characterise AI's domestic coming-into-being. Lastly, still in connection to the feminist relational premise, if we wish to situate the developmental context of AI in Thailand well and within its due nuance, we ought to also embed this review in the broader regional developmental context. For both latter points, given the limited scope of our paper, we may not provide an exhaustive discussion of either Thailand's computational history or AI's development in Asia and Southeast Asia. Though, we will therefore include key events which—we esteem—importantly influence Thailand's AI course of development.

The three expert interviews were conducted online, in semi-structured fashion. Professor Asanee Kawtrakul was interviewed on 20 February 2024, Professor Yuen Poovarawan on 10 May 2024, and Professor Wirote Aroonmanakun on 15 May 2024. All three are prominent figures in Thailand's AI landscape. Kawtrakul and Poovarawan worked on AI developments ever since AI's first steps in Thailand. Kawtrakul is a professor of computer engineering at Kasetsart University (Bangkok, Thailand) who worked on AI since the late 1980s. Poovarawan, recently retired, was a professor of computer science, also at Kasetsart University, who worked on AI since 1980. Kawtrakul and Poovarawan domestically pioneered research in natural language processing (NLP). NLP works on machine learning technology, a subfield of AI. Thai NLP developments were one of Thailand's key focus areas at the time when AI, as a 'foreign' technology, was being adopted domestically [2, 3]. Namely, the computational translation to the Thai language of software tools that were originally designed in the English language was key if willing to have a systemic incorporation of those software tools in Thailand. Poovarawan wrote the first textbook in Thai on NLP [13]. Aroonmanakun, a professor of linguistics at Chulalongkorn University (Bangkok, Thailand), joined this field of research in the late 1990s. He developed a Thai algorithmic word segmentation engine in 2002 [14] and was a key figure in designing the development of the Thai National Corpus in 2007[3] [15]. To ensure a certain reading fluidity in our paper, we do not provide a

---

[1] Though no less scattered across the Web.

[2] For reference, see the publication years of our references.

[3] A corpus is "the language data collected according to certain criteria to represent a language under examination" [15]. The Thai corpus is still publicly available today. Originally, the purpose was to find out which Thai words occur with other Thai words in a sentence, and how frequently [16].

verbatim transcription of the interviews; rather, we discuss them alongside the literature review.

Upon stating that research on AI's development in Thailand is scarce, not only does it reflect in the English-language literature (with 1 existing publication by Kawtrakul and Praneetpolgrang [1]). It also reflects in the Thai-language literature. Per 7 May 2024, Thai Journal Online, a portal of Thai language academic journals, returns 91 articles under the search terms "ปัญญาประดิษฐ์" (artificial intelligence) and "ไทย" (Thai). None of these articles discusses the history of AI in Thailand. Most are about various applications of AI: e.g., in education [17–19], banking [20]), agriculture [21], fact-checking [22, 23], and so on. Some articles regard Thai NLP technology (e.g., [24]). Some discuss AI and Thai law (e.g., copyright and liability [25, 26]). Lastly, a surprising number of articles focus on the acceptance of AI by various population groups (e.g., [27–30]. We will come back to this latter point later in our discussion.

3   A Brief History of AI in Thailand

Having discussed our aim and approach, we may now start our journey into AI's historical development in Thailand. As this description will reveal itself to be dense, to prevent your "asphyxiation" as reader [12], the timeline in Fig. 1 recapitulates key events characteristic of Thailand's AI development.

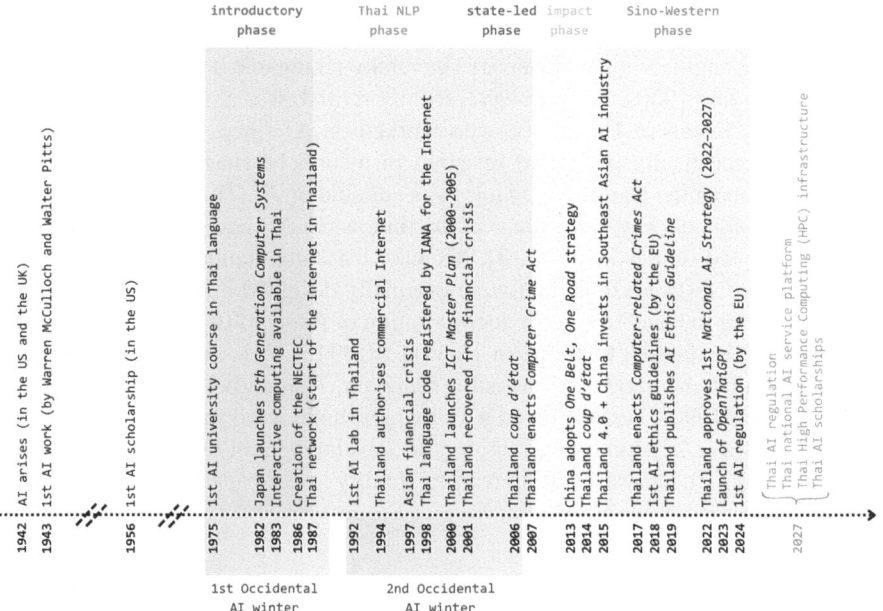

Fig. 1. Overview of key events characterising the historical development of AI in Thailand (timeline is not at scale; made with Canva).

3.1 1975–1987: The AI Introduction Phase

The Occident's influence over AI is undeniable. As already noted, AI was developed in the Western world in the early 1940s; allegedly rising simultaneously in the US and the UK in 1942 [4]. In 1956, it became a standalone scholarship in the US [5]. Japan too was experiencing an AI boom "in the late 1950s, just after the term "artificial intelligence" was coined by a computer scientist John McCarthy in 1956" [31]. This fact alone already reveals how presenting AI in Thailand as a technology transfer from the Occident in the 1970s is therefore a precipitated reduction of AI's 'thicker' and 'multiple' existence; of AI's more-than-Western existence [12].

It is no surprise that ideologies and regimes affect the way a society develops and, hence, the way a society develops scientifically and technologically. In that sense, periods *pre*dating the formal introduction of AI in a country too are informative if one is willing to gain an insightful appreciation of how AI developed across countries and regions[4]. Looking into the past of a contemporary major Asian player, we understand that "China held a critical and negative stance towards AI, likely influenced by the scientific and technological advancements of the former Soviet Union" [32]. At that time, China was therefore formally inactive in AI research and had to wait until "a campaign of ideological liberation began in the late 1970s, [for the revival of the] country's scientific community [and its ability to conduct AI research] in a legitimate manner" [32].

The West was going through its first AI winter from the early 1970s to the early/mid 1980s; a negative attitude emerged in the West vis-à-vis AI and funding steeply decreased as research failed to meet its promises[5] [33, 34]. It is during this first Occidental AI winter that AI was formally introduced to Thailand. The first Thai-language university course notes on AI appeared in 1975 [1]. However, even though AI was taught at Thai universities [35],

> "In the early 1980s, Thailand was a barren landscape for computer networking. Copper telephone lines were mapped thinly. [...] Computers were visible but not widely used. They were expensive and incomprehensible for most Thais, due to the lack of Thai language software and the lack of computer standardization" [36].

Indeed, even though "[i]nteractive computing came to Thailand in 1983" [2], "as a result of competition and corporate strategy to retain [...] customers[, by] 1984, there were at least 20 different character coding conventions used in Thailand. [...] the first standard Thai code for computers was developed in 1986[, thereby improving] the computer processing speed" [2].

Needless to say that computational developments in Thailand were effortful at least until 1986. Furthermore, Thailand had to wait until 1987 to benefit from Australian technical and financial assistance to develop its domestic Internet [36, 42]. Thailand's first inter-university UUCP connection was established. This network was "mainly intended for educational purposes[;] commercial usage was banned" [49]. This may partly explain why it took Thai scholars until the late 1980s to themselves innovate AI technology rather

---

[4] Given the scope of this paper, we do not discuss the periods predating the formal introduction of AI in Thailand, but it is a most interesting avenue to keep in mind for future research.

[5] There is no strict consensus around the dates defining the so-called AI winters.

than improve Western developments[6] [1, 3]. In fact, another factor explaining why Thai scholars started to innovate AI technology only in the late 1980s, was the creation in 1986 of the National Electronics and Computer Technology Center (NECTEC). As a research organisation part of the government's National Science and Technology Development Agency, the NECTEC served among other things to stimulate academic interest for AI [1, 16].

While the West was showing frustration toward unmet expectations, and while Thailand was in its early introductory phase, Japan underwent a time of great AI expansion. Firstly, between 1970 and 1973, Japan produced the first ever humanoid, the WABOT-1, using AI for its conversation system [37]. Secondly, the 1980s are commonly referred to as Japan's second AI boom [31, 38]. In 1982, in the frame of the *Fifth Generation Computer Systems* project, the Japanese government substantially funded AI until 1990; in particular, to develop expert systems[7] [39]. Noteworthily, "[i]n contrast to the Western projects [...], the Fifth Generation displayed many of the characteristics now described as essential to ethical, socially responsible AI" [38]. That a major Asian country had such influence over the AI industry is again suggestive of the precipitated and reductive character of the idea according to which the Western vision of AI alone would have determined the development of AI technology in Thailand (or internationally). Clearly, a more complex assemblage constitutes this domestic development. Indeed, as Poovarawan explained, during this period, Japan invested significantly in Thailand aiming to make it a research partner[8] [35]. A focal research area was machine translation.

3.2 1987–2001: The Advent of the Internet and the Thai NLP Phase

The end of the 1980s was synonymous with a general bonanza in the AI industry; "boom[ing] from a few million dollars in 1980 to billions of dollars in 1988, including hundreds of companies building expert systems, vision systems, robots, and software and hardware specialized for these purposes" [40]. Yet, despite this boom, the West entered its second AI winter from the late 1980s or early 1990s to roughly 2006 [33, 34]. Whereas the first AI winter was more about a failure on the research side, the second AI winter was more about a failure on the corporate side; "many companies [had] failed to deliver on extravagant promises" [40].

This Occident's second AI winter coincides with Thailand's "pioneering phase" [1]. Around this time, Thai AI scholars graduated from abroad—mainly from Edinburgh (UK), Japan, and the US—and set up AI research facilities in Thailand. With help from

---

[6] For instance, one such innovation was the need for teaching Thai word segmentation to the computer, because Thai language contains no spaces (a space for instance is like a dot in the English language) [35].

[7] Expert systems are "collections of rules which assume that human intelligence can be formalized and reconstructed in a top-down approach as a series of "if-then" statements" [4]. They are AI software of which the output makes the decision for the human; e.g., medical programmes that calculate the probability of an illness, or chess playing programmes that calculate the next best move.

[8] Poovarawan for instance received a fellowship in 1980 to start research on Thai NLP in Japan. The fellowship host, the Museum of Ethnology, was interested in processing ancient Thai texts [35].

the NECTEC, the first AI lab was created in 1992 at Kasetsart University and originally focused on NLP and expert systems[9] [1, 3, 35].

Thankfully to international cooperation, 1992 was also the year "that the Thai language was properly specified in the international standard ISO 10646" [42]. Not only did "[e]arly IT standards in Thailand [define] the code points for computer handling, keyboard layout, and [Input/Output] method" [2]. It also provided the vital support necessary for the gradual transition of the Thai academic network "to full TCP/IP"[10] [36]. To compare with a Southeast Asian country that later became a global AI leader, Singapore had its first Internet in 1990 and established its first domestic Internet Service Provider (ISP) in 1991 [43]. In Thailand, the Communications Authority of Thailand only approved the commercialisation of the Internet in 1994 [42]. From 1994, the use of the Internet was thus no longer secluded to the academic realm and could reach the general Thai population [36, 49]. By providing funds in 1995 to build a line from Thailand to Japan to expand Thailand's academic network, Japan figures again as an important international aid to Thailand[11] [36].

Japan's influence over Thailand is perhaps more penetrating than just financial aid. We have already mentioned Japan's *Fifth Generation Computer Systems* project running from 1982 until 1990. Whereby we mentioned that the thought that the Western vision of AI would have determined the development of AI technology in Thailand may sound precipitous and reductive of a more complex assemblage. For while the West was undergoing its second AI winter, Colin Garvey insightfully describes how Japan was able to instil a new domestic sociotechnical vision or 'imaginary' [44]:

"[The Fifth Generation] committee's vision for the Japanese society of the 1990s involved avoiding the mistakes of its modernized predecessors. Specifically, whereas the United States and the United Kingdom had responded to social ills through neoliberal reforms, entrusting their cure to the power of markets, Japan would avoid "falling into their rut" by steering the domestic computing industry toward the production of socially beneficial computers—machines that would more evenly distribute the potential benefits of computing to ordinary people [...]. In the Japanese information society of the 1990s, computing would be a state-sponsored public good, rather than intellectual property to be patented and licensed to a set of domestic firms, [...] or spun off at public cost to private actors in what economist Mariana Mazzucato has called the "socialization of risk and privatization of rewards" process distinctive of Western neoliberal regimes" [38].

Given Japan's generous investment in Thailand at that time [16, 35], as well as the fact that many Thai scholars studied AI in Japan [1], indubitably, Japan's vision influenced

---

[9] Expert systems focused on agricultural and environmental systems [1].

[10] A necessary network transition to enable a connection to the international network, since TCP/IP is the most widely used Internet protocol.

[11] Thailand was also the provider of such regional aid. In 1995, and through further cooperation with Australia, the Thai Social/Scientific Academic and Research Network "sponsored Laos Academic Internet Demonstration [and later morphed] into a formal grant aid by [the] Thai government to [the] Laos government" [42].

how Thailand envisioned its domestic adoption of AI in the 1990s[12]. It may also have been of influence on Singapore, which was also undergoing its AI expansion phase in the 1990s[13] [45].

The gradually improving Internet infrastructure in the 1990s certainly contributed to the increase in Thailand's import of personal computers[14]; estimated at 680,000 in 1994 and at one million in 1996 [46]. In 1996, the first Thai-US fibre optic Internet connection was also built [42]; "about 10 per cent of [Thailand's] personal computers [were] connected to the Internet" [46]. This increase created a high demand for the effective use of Thai language on the personal computer that was of English-language design. Linguists therefore joined the efforts of computer scientists. In effect, Thai NLP technology was the first domestic AI research project, also leading to the first tangible AI research collaboration. The pioneering work in the development of Thai NLP technology led to the creation of Thai inter-university collaborations between computer scientists from Kasetsart University and linguists from Chulalongkorn University, and led to the creation of the first Thai Symposium on Natural Language Processing, which involved [16, 35, 41].

It is worth noting that while the Thai computational language was already conforming to the international standard ISO since 1992 [42], due to procedural obstacles met in the process of having the Thai code meeting the Latin code[15], the Thai code was only "formally registered by the Internet Assigned Numbers Authority [as] a legitimate encoding for the Internet in 1998" [2]. While beneficial in multiple respects, processes of standardisation are also famed impediments; and this case illustrates this very well.

With the improving Internet infrastructure, the Thai academic network that accounted for the majority of domestic Internet users also expanded. More AI labs and institutes started to emerge, in turn, widening Thailand's AI research horizon. Speech processing, NLP, and expert systems were the priority research areas within academia; image processing was a secondary, less trending area of academic research[16]; and expert systems were almost the sole focus of development in the private sector [3].

---

[12] Wirote Aroonmanakun mentioned that around this time he had also joined a project on Thai NLP supported by the French government [16].

[13] Japan's Fifth Generation project would have failed when the US came up with cheap personal computers, because the Japanese hardware was too slow to be commercially viable [35].

[14] As an interesting sidenote: in the early 1970s, Chulalongkorn University had one big computer owned by the American military who used it for their activities during the Vietnam War; Thais were not allowed to use it [35].

[15] See the publication by Hugh Thaweesak Koanantakool, Theppitak Karoonboonyanan, and Chai Wutiwiwatchai for a very well-written explanation of the difficulties met in the process of encoding the Thai language to meeting prior established international standards [2]. Moreover, as Poovarawan explained, the computation of the Thai language had to abide by the Thai language rules as laid out by the Thai state body governing the use of Thai language, the Royal Society of Thailand [35].

[16] Other areas of research were: "machine learning, robotics and other mechanical instruments, intelligent computer-aided instruction [. . .], intelligent information retrieval, information extraction and summarization, data mining and knowledge engineering, intelligent control systems, and forecasting systems" [1].

While Thailand had "announced its second [...] International link" in 1997 [42], it was also the year that Thailand was hit by a financial crisis; commonly referred to as the *1997 Asian financial crisis*. Allegedly, because it implemented an economic liberalisation too wildly so to speak [47]. The sixteen Thai ISPs that serviced 250,000 to 350,000 Thai users in 1998 "were severely hit by the lower local demand and the baht[17] devaluation" [36, 46]. In 1999, "the Internet user base in Thailand [was] only about 0.7% of the total population" [36]. As is known, the breadth of Internet usage in a country affects that population's digital literacy, in turn, affecting that country's readiness for the adoption of new digital developments.

3.3 2001–2007: Post-Financial Crisis Times and the Government-Led Orientation Phase

Thailand recovered from its financial crisis in 2001 [47]. By 2002, "almost all households had electricity and [...] the latest technologies such as the mobile phones and the internet had to varying degrees been incorporated into village life (Phatharathananunth, 2016)" [48]. Whereas the Internet usage re-expanded, the Ministry of Information and Communication Technology[18] clamped down, launching a "campaign to 'clean up cyberspace, in terms of both morality and national security'" which led to website blockages also in relation to the criminal act of *lèse majesté*[19] [49]. That government practices orient a population's perception of that government is nothing new. As put by Sheila Jasanoff, Ian McGonigle, and Hallam Stevens, the "design [of] the social, political, and educational systems in which [technologies] are embedded [shape] how we [...] live with technology" [50]. By enacting strict laws, a government shapes its population; it shapes the population's perception of the state, in turn, affecting the country's 'sociotechnical imaginary' [44].

This shift whereby the government took more control over various aspects of Thai sociotechnical life also reflects in its approach to AI. One may in fact speculate that as the 1997 Asian financial crisis brought criticism onto the former "loose" approach to market regulation, it thereby led the government to take more affirmative lead in its country affairs. In 2000, as part of the government's *Information & Communication Technology Master Plan (2000–2005)*, the National Research Council of Thailand created research road maps to maximise the effectiveness of public money allocation [1, 41]. While still recovering from the crisis and while seeking to shape the Kingdom into a digital knowledge society to be part of the global digitisation trend, domestic AI research now became primarily oriented by the government [41]. The aim was to digitalise five sectors: the government, education, society, the industry, and commerce. AI was regarded as a major driver and facilitator for the plan's successful completion. NLP systems were again of particular importance; for instance, for "word [segmentation], [optical character recognition], handwriting recognition, text-to-speech and speech synthesis[,] voice recognition, machine translation, and search algorithms" [2]. Given the accentuated

---

[17] Baht is the Thai currency.

[18] Known today as the Ministry of Digital Economy and Society.

[19] Whereby "acts against the king [are] acts against the state" [49].

need for interdisciplinary collaboration, more systematic collaborations between Thai research centres and Thai universities arose.

In 2006, Thailand underwent a *coup d'état* and, in 2007, it enacted its *Computer Crime Act* which has been criticised by some for its censoring potential [49]. We have already expressed how government practices orient a population's perception of that government, and how this in turn affects a population's adoption of technologies that this population might therefore not trust. It is further important to note that government practices can have diplomatic repercussions and orient international aid; in turn, influencing a country's domestic technoscientific trajectory (we will come back to these points later).

3.4 2007–2015: The Impact-Based Development Phase

In the continued pursuit of shaping the Kingdom into a digital knowledge society, and in the continuity of its 2000–2005 master plan, research remained "determined by the national development plan" [1]. However, what changed was that research was now assessed through quantitative societal impact indicators, like generated "incomes, publications, prototypes, and patents" [1]. As such, AI research turned itself yet more toward practical, tangible, and societally impactful developments; and yet less toward abstract or theoretical developments.

Healthcare became a key new AI research focus[20]; for instance, developing medical diagnostic systems (e.g., AI semen analysis for in-vitro-fertilisation). Another newcomer is the tourism industry (e.g., the creation of smart hotel rooms in luxury accommodation to offer a personalised experience to customers; [52]). Agriculture has remained a main focus since the 1990s (e.g., for rice quality inspection; [51]). NLP technology kept being an area of improvement as it was an important means for Thailand to extract knowledge from the digital space and to ensure Thai knowledge accumulation; which in turn served to ensure the perennity of its AI engineering activities. In a way, we may perhaps liken the use of NLP systems in this sense to that of Web crawlers.

While Thailand at this point was seeking societally impactful AI developments, in the 2010s, South Korea was rising to becoming a global AI leader [53]. China on its side was "in a period of exponential growth in AI [with a particular focus on] image classification, speech recognition, automated knowledge Q&A, human-machine chess, and automatic driving" [32]. When China adopted its notable *One Belt, One Road* strategy for international development in 2013, a year later, Thailand underwent another *coup d'état*. As mentioned before, foreign aid is importantly affected by such events. For Thailand, the successive coups in 2006 and 2014 led the US government "to cut military aid to Thailand and [to] publicly [criticise] the military regimes" [54]. In this wake, because the US-Thailand relations "seemed headed for a more strained and distant relationship[, China in turn] has become more interested in deepening cooperation" [54]. Invariably, this influences Thailand's domestic technoscientific trajectory—at least to some degree.

---

[20] Thailand's focus on healthcare reflects today in its multiple JCI-accredited hospitals, and in Bangkok having grown to the status of Asian medical hub, thereby attracting medical tourists.

3.5   2015–2024: Thailand 4.0 and the Sino-Western Phase

2015 was the year introducing the notion of the *4th industrial revolution*, emphasising how the latest technologies (among which AI) occasion a disruption from the previous industrial revolution, the *information age*, itself characterised by the advent of information technologies (e.g., the personal computer). Even though "by the year 2015, only 40 percent of the Thai population had access to the Internet" [48], Thailand too, like most countries, exhibited keen ambition to join this 4th revolution:

> "[the Kingdom's policymakers have devised] an economic model based on creativity, innovation, new technology and high-level services. The aim [...] is to metamorphose the Kingdom into a value-based economy by reforming its existing major industry clusters (i.e., automotive, electronics, affluent medical and wellness tourism, food, agriculture and biotechnology) and scaling up the development of new sectors such as robotics, digital industry, aviation and logistics, biofuels and biochemical, as well as further solidifying Thailand as a major regional medical hub" [55].

Within the framework of Thailand 4.0, AI technology features as one of the core factors supposed to ensure its 4.0 revolution [9, 55]. The government dedicates particular attention to the sustainable development of AI; striving for social equality, economic competition, green growth, and national stability[21] [55].

This strive for what we may denote as more 'noble' goals besides the materialistic goal of wealth accumulation was also an observation made by Kawtrakul [41]. She expressed that Thailand is very aware of the need for ethical AI. This has resulted in a number of domestic AI ethics guidelines. In 2019, the Thai government approved the *Digital Thailand – AI Ethics Guideline* [56]. The guidelines are not legally binding, but it represents the first step toward an ethics of AI in the country. In 2022, Chulalongkorn University published its own *Thailand Artificial Intelligence Guideline 1.0* which contains key principles and recommendations for the responsible development of AI technology and a risk assessment framework [57]. The National Science and Technology Development Agency too published an AI ethics guideline in 2022, mostly for internal use [58]. Per the year 2022, Thailand would comprise no fewer than ten AI ethics guidelines [59]. To Elina Noor and Mark B. Manantan:

> "Thailand's promulgation of [...] AI ethical guidelines signals its interest and willingness to join the ranks of highly advanced countries aiming to become fair and equitable data-driven economies" [8].

---

[21] As a sidenote, while the sustainable development of AI is a goal formally shared among all UN member states, their contextual application differs significantly. For instance, in relation to the potential impact of AI on Thailand's workforce, the ESCAP notes in a 2018 report that, "what is technically feasible is not always economically viable", and that for Thailand an "initial investment on robots cannot be recovered within the 15 years life span of the machines [...] while the payback time in Republic of Korea, Japan, New Zealand, Singapore and Australia can be only around 1.5 years or less" [76].

Codes for the ethical and responsible development of technology have long existed. In the world of Information and Communication Technology, the International Federation for Information Processing for instance published an updated *Code of Ethics and Professional Conduct* in 2021 [60]. Looking into the codes that inspired Thailand would be an interesting avenue of further research to understand which ideologies infused Thai's AI development (e.g., the global impact of the EU's AI Act for instance may not be omitted; see also Soraj Hongladarom and Jerd Bandasak for an interesting analysis on how non/western AI ethics guidelines compare [61]).

Another way in which Thailand's attention for ethical AI comes forth is the launch in 2020 of the international Feminist AI Research Network, f < a + i > r. Chulalongkorn University has been a partner since its launch. This network might not be government-led, but it is indicative of a certain academic sensitivity for "the plural [micro]histories of AI" [62]. A sensitivity toward the 'thickness' of AI and AI in Thailand [12]. "The fact that [feminist academics, students, NGOs, and artists in the global South and North] are getting involved in rethinking AI brings a different flavor from the conventional history of AI" [62]. In that vein, a noteworthy number of publications from 2019 onward seem to focus on the acceptance of AI by various population groups. Namely, when we exemplified existing Thai literature on AI in our methodology, all these publications were from 2018 and later. Even though it was a specified keyword search on "AI" and "Thai", it is in a way indicative of a certain willingness to reach a harmonious vision or 'imaginary' of AI within the Thai digital knowledge society. At least, among the Thai academic body.

This pursuit for the more 'noble' could among other factors be a reaction to the "increasing awareness among Thais about their digital rights, plus growing concerns about state surveillance and discrimination" [8]. In 2017, Thailand enacted an amended version of the 2007 *Computer Crime Act*: the *Computer-related Crimes Act*. While offering greater protection of Thai society, the new act has still been criticised for lacking "full compliance with international human rights standards" and allowing "the government nearly unfettered authority to […] engage in surveillance" [63]. Therefore, as expressed by Noor and Manantan:

> "Thailand's AI narrative can no longer just rely on its digital economic success. The focus on AI's transformative power is shifting to its real-world causes and effects as well as how those will, in turn, impact the social fabric of Thailand's emerging digital society. [With] a trust deficit is looming over the government's plans for a Smart Thailand[, mustering] public support will be crucial to the continuing viability of Thailand 4.0" [8].

In July 2022, the NECTEC approved the *(Draft) Thailand National AI Strategy and Action Plan (2022–2027)*[22] [64]. The development of socially impactful and sustainable AI are part of the plan's five focus strategies. Among the goals to reach by 2028 are that of having "[a]t least 600,000 Thai population have awareness of AI law and ethics

---

[22] This plan follows the preceding *Thailand Digital Government Development Plan* (2017–2021), in which AI already figured, but where AI was far from being the main sociotechnical phenomenon around which the development plan was shaped (see [65]).

[as well as to have an enforced] AI Law & Regulation" [64]. Still within this strategy, and in continuation of the early footsteps of Thai NLP, Thai computational linguistics remains a key focal area, for instance, to develop Thai large language models such as Typhoon (see [66]). Like many countries, Thailand therefore also grew significant interest in machine learning technology particularly from the mid-2010s onward [41]. Another Thai large language model development has been the release of OpenThaiGPT in 2023. This project is mainly a collaboration between Thailand's *Artificial Intelligence Entrepreneur Association* and its *Artificial Intelligence Association*, which also received support from many international and domestic organisations.

As a matter of comparison with regional competitors, Singapore launched its first national AI strategy in 2019, and an amended strategy 2.0 in 2023 [9, 67]. Indonesia did so in 2020 and Malaysia in 2021 [8]. Generally speaking, while "Southeast Asia is still in the early stages of AI adoption" [9], it is progressively rising from obscurity; albeit at different internal speeds. Singapore is indeed a prominent player in the AI field; not only as a regional "frontrunner in AI experimentation in financial services, high-tech telecommunications, manufacturing, and mobility" [8], but also as an international leader [68].

Today, the US generally remains indexed as a global leader in the AI industry. Yet, China's recent rise is distinct (e.g., see [10]). Since the 2010s, China "witnessed explosive growth" in its AI industry [68]. With its 730 million Internet users, China naturally owns a huge data well to enable the necessary sophistication of its AI systems [69]. In the wake of China's 2013 *One Belt, One Road strategy*, and because Western countries are "uneasy with the presence of Chinese tech companies, [...] Beijing has been gradually shifting its focus to other regions, particularly Southeast Asia" [68]. Indeed, in its race to AI leadership whereby China established AI as a top priority in its 2015 plan *Made in China 2025* [68], China's investment in Southeast Asia's AI industry almost exponentially increased between 2017 and 2022 [68]. While Singapore and Malaysia feature among China's current favourites:

"China has [also] established technology transfer centers in Cambodia, Indonesia, Myanmar, Laos, and Thailand. These transfer centers act as forums for collaboration between companies, research institutions, and other actors to facilitate the dissemination of technology across ASEAN from China" [68].

China's AI focus being on "cloud computing, big data, the Internet, and the Internet of Things, [and] computing platforms" [32], one may imagine this to influence the course of AI's industry development in Thailand. For instance, an ongoing Sino-Thai collaboration is that of developing Thailand's first metaverse [68]. Furthermore, given China's active engagement in AI policy (see [70]), the way that China regulates AI will certainly reflect in some ways in Thailand's AI policies (cf. Developing an AI regulation is part of Thailand's 2022–2027 AI strategy). Other countries too remain investment partners in Thailand's AI industry and therefore influence AI's domestic course of development (e.g., Japan's *Science and Technology in Society forum* that also convenes in Thailand to discuss how society would need to shape to enable scientific and technological innovation). What is certain is that the AI tides are gradually shifting, potentially positioning Southeast Asia as part of the leading AI ranks in the near future.

We may speculate that the reason for China to focus on Southeast Asia rather than East Asia—the latter which is comparatively way ahead of Southeast Asia[23]—may find a simple explanation again in international relations and in the known US-China tensions. The US for instance is an important investment partner of South Korea [53]. For Taiwan, given China's attempts at making Taiwan its own, Taiwan therefore avoids partnering with China. As for Japan, it has been a key provider of foreign investment to the US for some time (see [71]). Hence, we coined this phase the *Sino-Western phase*, because China's influence is indubitable. Yet, the Occident remains influential, albeit, in less visible ways (e.g., since a majority of AI software is designed in the English language, still today, Western developments cause Thai AI developments to be importantly focused on Thai NLP systems [2, 16, 35]). We may indeed wish to highlight that despite China's significant involvement in Southeast Asia, it does not efface Western influence altogether. Major Western companies, for instance, are ongoingly investing in the region; "Microsoft [...] is building new cloud and AI infrastructure [in] Indonesia and Malaysia[, committing] $1.7 billion and $2.2 billion respectively to advance AI efforts. [...] In Thailand, Microsoft will build its first data center [and] and pledged to train over 100,000 people" [72]. So too, international organisations that are reminiscent of Western humanitarian ideologies remain in the loop when it comes to the Southeast Asian AI industry. The UN's *Economic and Social Commission for Asia and the Pacific* (ESCAP) for instance partners with Thailand since 2019 to ensure the sustainable development of AI [73]. The ESCAP's AI development focus in Thailand is particularly aimed at healthcare and medicine, and poverty alleviation.

4 Speculated Thai AI Futures

To Kawtrakul, Thailand offers many promising areas of research in AI; such as robotics, image processing, and optical character recognition [41] (see also [9]). In 2017, Will Baxter wrote that, "in terms of broader implementation of automation and digitalization, Thai industry to date has unfortunately leaned more heavily toward inertia rather than initiative" [55]. This is also a view that we found again in Deloitte's report on Thailand's digital transformation [7]. Perhaps this is where Kawtrakul makes a good point in suggesting that a matter that requires dire attention is the development of a concerted effort to create a central repository cartographing AI research in Thailand. Through such open-access knowledge database, technology readiness levels would be easier to identify and AI developments could be better matched with priority societal challenges. To Kawtrakul, this would not only benefit research, but also evidence-based AI policy-making, in turn, better supporting the creation of a thriving Thai digital knowledge society [41]. By making the creation of a "national AI service platform" one of its priorities within its 2022–2027 AI strategy, the Thai government exhibits awareness and action-taking to deal with this issue [64]. We may also wish to highlight that, whereas ample research

---

[23] Japan, Taiwan, and South Korea are always ranking among top AI leaders, whereas in Southeast Asia, currently speaking, only Singapore is similarly indexed.

discusses how AI technologies for instance contribute to the reification of socioeconomic inequalities[24], the lack of such a central repository for Thailand thereby impedes an adequate scrutiny of the societal effects of AI on the situated sociotechnical context of Thailand.

Another site fit for optimisation to Kawtrakul and Aroonmanakun, are public-private collaborations [16, 41]. From Kawtrakul's interview, we may understand that government funding for AI research at private Thai companies and research labs remains insufficient. Kawtrakul's logic is that since those private entities already have the necessary knowhow and equipment, this would be an efficient allocation of public resources (cf. Thailand's interest for societally impactful AI developments). Supporting the creation of such synergistic public-private dynamic would (i) benefit the creation of a trusting public-private relationship, (ii) benefit the creation of commercial opportunities, and (iii) support the collective construction of a thriving Thai digital knowledge society. While Thai entrepreneurs were already taking the lead on this, the Thai government is once more very aware of this lacuna. A key focus of the 2022–2027 AI strategy is to promote more integration of AI technology within Thailand's various sectors. Main targets for AI integration are: the government (to have the agencies use AI tools for its services and for public surveillance), education (to educate about AI and to personalise education through AI), healthcare, farming, the financial sector, and the automotive industry [64]. Part of the government's areas of focus is its awareness of Thailand's ageing population. Perhaps a more personal addition from Kawtrakul would be the need for AI applications to deal with damaged roads, abused young children, and drug addiction among teenagers [41].

Clearly, in all these planned domestic AI developments for the near-future, Thai computational linguistics will remain a key component of Thailand's AI strategy. Still today, "[m]ost of the available software tools for developing information retrieval [systems] have not been specifically designed to work with Thai" [2]. Therefore, still today, the early footsteps of AI in Thailand—then predominantly focused on Thai NLP—show to be of immense importance in ensuring a smooth domestic AI strategy. As was wisely noted by Hugh Thaweesak Koanantakool, Theppitak Karoonboonyanan, and Chai Wutiwiwatchai in 2009:

> "Because computers are also becoming indispensable for the elderly, [for] persons with disabilities, and [for] those who are illiterate, many new developments will be possible. Speech [Input/Output], handwriting input, sign languages, mobile-phone text entry, a voice-command system, speech-to-speech translation, voice search, and so on are the most likely efforts to be developed and commercialized" [2].

Not only does this exhibit how Thai academics have a sensibility toward seeking an inclusion of the Thai population in all its diversity and 'thickness' into the contemporary AI phenomenon. It also reflects that however much Thailand's domestic AI development may be undergoing foreign influence (be it because of the technical origins

---

[24] For instance, Kelly Joyce and colleagues provide a rich discussion on the societal impact of AI [74].

of AI or because of foreign aid programmes or because of competing foreign technology developments), to a certain extent, Thailand will always shape AI as its own.

Speaking of foreign influence, given the gradual shift from what was originally a predominantly Western influence[25] to a Sino-Western influence on AI's development in Thailand, and given that Thailand is nowhere near to becoming a global AI leader, the developmental course that AI will follow will certainly depend on China's AI strategy and perhaps will to a comparatively lesser extent depend on other foreign influences. Like most countries around the world, China too exhibits a characteristic fondness for surveillance. Yet, different countries have different political contexts. Hence, alongside China's fondness for surveillance, with Thailand's self-declared interest in AI for public surveillance [64], we may expect further Thai AI developments in facial recognition and smart policing technology (see also [75]). While it is certainly so that AI for public surveillance figures as only just one point among multiple points of interest to the NECTEC, given the situated occurrence of this interest, as mentioned by several organisations and scholars, this interest ought to receive all its due attention [8, 63, 75]. In this context, if the Thai government wishes to have a smooth unfolding of its AI strategy, it will want to avoid a "trust deficit" within its population [76]. Thence, in pursuit of such heightened public acceptance of its strategy, we could speculate on more particular Thai developments in eXplainable AI (XAI) (see Fatemeh Alizadeh, Gunnar Stevens, and Margarita Esau for a comprehensive discussion on lay perceptions of AI and how this affects trust [77]). Simultaneously, through international non-governmental initiatives such as that of $f < a + i > r$, we may also speculate the birth of more non-government-led efforts in developing Feminist AI (FAI) (see [62]).

5 Conclusion

AI is veritably inundating our everyday lives. With the monstrous number of news articles, podcasts, research publications, scholarships, and so on devoted to AI, and the mushrooming number of AI applications into mundane everyday activities, it is fair to say that AI is among the most popular technologies of contemporary history, and hence why countries around the world are developing national AI strategies. Tracing AI's history in Thailand therefore makes all the more sense should we wish to craft a better understanding of Thailand's contemporary sociotechnical unfoldment.

The first steps of AI in Thailand are traced back to Thai university course material from 1975. Academia played a crucial role in introducing AI in Thailand; and it continues to play a crucial role to this day. However, a critique we might express vis-à-vis academia is its lack of historical and sociological interest in AI. The literature on the history of AI in Thailand is practically a wasteland. There have been qualitative surveys in 1992 and 1999 on the use of AI [3]; and in 2014 appeared the first publication tracing the history of AI in Thailand [1]. Since about 2015, multiple reports can be found on Thailand's digital transformation; yet, these always only adopt the traditional economy lens, only interested in comparatively indexing countries worldwide based on productivity performances of domestic AI industries. This paper sought to fill in this current literature gap through

---

[25] Yet, in many ways, this influence was also importantly characterized by Japanese aid.

offering a centralised compilation of information about the historical unfolding of AI technology in Thailand from its beginnings to this day, to facilitate future research. Given the scarcity in written forms of publicly available knowledge, and to better weave together the relations that characterise AI's domestic coming-into-being, we enriched the available literature in three ways: (i) by conducting three qualitative expert interviews, (ii) by including—what we saw as—key events in Thailand's history of computation and the Internet, and (iii) by including—what we saw as—key regional players that influence(d) the development of AI in Thailand.

Thailand is regularly overlooked or marginalised when we 'think AI' [12]. Precipitous understandings of AI might speak of a technology transfer of the 1970s from the Occident to Thailand. However, through compiling information on the development of AI in Thailand, we could savour the nuances that make this tale more sapid and flavourful. Despite our unavoidable situatedness in compiling this tale [12], as well as the therefore unavoidable absences from our tale[26], through dedicating careful attention to the contextual development of AI in what is an often-marginalised country in this respect, we could reveal two things. Firstly, it revealed the complexity of AI's relationally interwoven unfolding (i.e., AI's thickness at domestic level). Secondly, it revealed that in order to have an understanding of the historical development of AI across countries, it is in fact *compulsory* to understand the development of AI in Thailand[27] (i.e., AI's thickness at global level). In other words, Thailand may not be skipped if one wishes to craft a comprehensive understanding of how AI developed in general.

Firstly, in terms of the revealed relational thickness of Thailand's domestic development of AI, linguistics clearly play(ed) a major role. Thailand's historical development in computational linguistics therefore too proved to be a key factor especially in terms of smoothening this unfoldment. The infrastructural context and related economic costs also greatly help explain AI's characteristic sociotechnical development in Thailand (e.g., how a lack of financial means domestically causes a reliance on foreign aid, or how the availability of and access to technology affects the Thai population's technical literacy and therefore its readiness to adopt new technology). This is also the part where the history of the Internet in Thailand entered the scene, as well as some political and regulatory events that oriented AI's development by way of orienting domestic perceptions of AI and foreign investment interests (e.g., the two consecutive coups that put the US off and led to a heightened interest from China). Early investment partners that enabled Thailand's AI development were Australia and Japan[28]. Furthermore, as many Thai scholars studied AI abroad—notably so in Japan—Japan figures as an influential actor also in terms of instilling a certain 'vision' of AI among the Thai academic body

---

[26] In the sense that we unavoidably leave out some elements part of this tale; whether advertently due to the scope of this paper, or inadvertently due to a lack in our understanding.

[27] Cf. The feminist premise according to which phenomena come into being through relations that are in constant becoming. Meaning that phenomena, or parts thereof, cannot be cut out as if unrelated to that phenomenon or others. In this sense, what happens in Thailand's AI landscape cannot be secluded from what happens in other AI landscapes, and vice versa.

[28] France too would have been an early investment partner for developments in Thai NLP [16], though, the amount of French investment in Thailand seems to have been relatively small compared to that of the other countries discussed.

(alongside the visions of AI as transmitted from scholarships in the UK and the US). The fact that AI was early on a more-than-Western technology certainly helped the fact that while the West underwent its two AI winters, Thailand could still find foreign partnerships to develop its knowhow and infrastructure. With the Vietnam War, the US certainly also influenced early on how Thailand could develop its computer network and associated literacy (in essence, a restriction of this potential by disabling access to the Thais of then existing computational equipment; [35]). Further on, technological developments in the US that competed with Japan's (notably so the personal computer), thereby oriented which country it was that could exert foreign influence over Thailand's AI development. More recently, China's growing interest since 2015, and growing investment since 2017 [68], is a rather noteworthy feature of AI's development in Thailand.

Secondly, in terms of how our tale of AI in Thailand enables to thicken the tales of AI abroad, Thailand clearly plays an active role in the development of AI, also abroad. This contrasts significantly with the understanding according to which Thailand, as a non-leader, would purely be subjected to 'foreign' influence in the process of this technology transfer[29]. Foreign influences are not one-way directions; they are two-way relations. Firstly, during technology transfers, technological artifacts always end up being "used in ways entirely different from their designed purposes" [11]. Hence, in the process of adopting AI, Thailand necessarily makes it something of its own. For instance, with the Thai NLP efforts, or with the potential future development of a niche AI leadership (e.g., in the field of medicine given Thailand's ambition to become an international medical hub). Secondly, the very fact that Thailand *chose* to adopt AI (to whatever degree of foreign influence or 'aid'), reflects Thailand's 'self-interested' goal so to speak of attaining economic growth (if not academic growth too). Lastly, the very fact that Thailand adopted AI very much benefits foreign investors in their own self-interested pursuit of economic growth or diplomatic bargaining power. Therefore, the same way that what other countries do with AI influences how Thailand goes about AI, so too, what Thailand does with AI domestically influences how partnering countries go about AI (e.g., Thai AI innovations in facial recognition or medical applications for instance affect how other countries innovate in those fields, be it by innovating further on the Thai innovations)[30]. Thailand's role in AI's historical development is thus as important as are the roles of the countries we 'naturally' (normatively) tend to think about upon 'thinking AI'. That is the very reason for which Thailand's part in AI's history may neither be forgotten nor marginalised.

The way AI develops in Thailand is very much an expression of Thailand's way of relating to foreign technology and to the world; very much characterised by an academic curiosity and a governmental will for sociotechnical development. With this paper, we hope to insufflate new air into this musty research avenue for it to regain momentum.

---

[29] Which is an understanding that, in turn, would also abide by the Heideggerian perception of technology as a "weighty", subjecting entity that is outside of one's control (see [11]).

[30] Whether these dynamics amount to the same degrees of influence is certainly a matter of debate. One might for instance abide by the economic lens, stating that more developed countries surely can exert more influence over less developed countries. Yet, as history has shown over and again, economic power is not the sole ingredient to secure a stronghold over international sociotechnical influence.

Given our focus having been more so on the scientific adoption of AI, further research might be interested in taking a closer look at the Thai population. Ample research discusses how the development and application of certain AI technologies contributes to the reification of socioeconomic inequalities (see [74]). Yet, little is known in this sense for Thailand. Still in this vein, more attention could be dedicated to understand the Thais' perception *of* AI and their expectations *with* AI. A last suggestion in this sense could be to adopt a postphenomenological look at AI in Thailand (see [11]), such as to understand the situated phenomenological and heuristic peculiarities upon adopting what was originally a 'foreign' technology (and gradually became an *also* 'own' technology).

Acknowledgement. We were very fortunate that Professor Asanee Kawtrakul, Professor Yuen Poovarawan, and Professor Wirote Aroonmanakun agreed to talk with us and would like to express our special gratitude to them. We would also like to thank the reviewers for their very helpful feedback in shaping this paper as it is at present.

Funding. This research was partially funded by a grant from the National Research Council of Thailand.

Contributor Statement. Suppanat Sakprasert conducted the qualitative expert interviews. Soraj Hongladarom and Auriane van der Vaeren wrote the paper.

References

1. Kawtrakul, A., Praneetpolgrang, P.: A history of AI research and development in Thailand: three periods, three directions. AI Mag. **35**(2), 83–92 (2014). https://doi.org/10.1609/aimag. v35i2.2522
2. Koanantakool, H.T., Karoonboonyanan, T., Wutiwiwatchai, C.: Computers and the Thai language. IEEE Ann. Hist. Comput. **31**(1), 46–61 (2009). https://doi.org/10.1109/MAHC. 2009.5
3. Kijsirikul, B., Theeramunkong, T.: Survey on artificial intelligence technology in Thailand. Chulalongkorn University & Thammasat University, Bangkok (1999). http://107.167.189. 191/~boonserm/publication/AISurvey.pdf
4. Haenlein, M., Kaplan, A.: A brief history of artificial intelligence: On the past, present, and future of artificial intelligence. Calif. Manage. Rev. **61**(4), 1 (2019). https://doi.org/10.1177/ 0008125619864925
5. Moravec, H.: The great 1980s AI bubble: a review of the brain makers. AI Mag. **15**(3), 86–87 (1994). https://doi.org/10.1609/aimag.v15i3.1105
6. Wutiwiwatchai, C.: Thailand's AI strategy to boost economic and social wellbeing. OECD (2022). https://oecd.ai/en/wonk/thailand-ai-strategies
7. Chutijirawong, N., Hora, V., Bunsupaporn, K., Bunyalug, C., Satityathiwat, S.: The Thailand digital transformation survey report 2020. Deloitte (2020). https://www2.deloitte.com/con tent/dam/Deloitte/th/Documents/technology/th-tech-the-thailand-digital-transformation-rep ort.pdf
8. Noor, E., Manantan, M.B.: Data and artificial intelligence in Southeast Asia. The Asia Society (2022). https://asiasociety.org/policy-institute/raising-standards-data-ai-southeast-asia/ai/ thailand
9. Mongkol, K.: The future of artificial intelligence in Southeast Asia: the case of Thailand. In: Ordóñez de Pablos, P., Zhang, X., Almunawar, M.N. (eds.) Handbook of research on artificial intelligence and knowledge management in Asia's digital economy, pp. 12–35. IGI Global (2023). https://doi.org/10.4018/978-1-6684-5849-5.ch002

10. Stanford Institute for Human-Centered Artificial Intelligence: Artificial intelligence index report 2024. Stanford University (2024). https://aiindex.stanford.edu/report/
11. Ihde, D.: Postphenomenology: Essays in the Postmodern Context. Northwestern University Press, Illinois (1993)
12. Puig de la Bellacasa, M.: 'Nothing comes without TTS world': thinking with care. Sociol. Rev. **60**(2), 197–216 (2012). https://doi.org/10.1111/j.1467-954X.2012.02070.x
13. Poovarawan, Y., Wongchaisuwat, C.: Natural language processing [การประมวลผลภาษาธรรม ชาติ]. King Mongkut institute of technology, Thonburi (1992)
14. Aroonmanakun, W.: Collocation and Thai word segmentation. In: Proceedings of the Fifth Symposium on Natural Language Processing (SNLP) and the Fifth Oriental COCOSDA Workshop. Sirindhorn International Institute of Technology, Pathumthani, pp. 68–75 (2002)
15. Aroonmanakun, W.: Creating the Thai national corpus. Manusya J. Humanit. **10**(13) (2007)
16. Aroonmanakun, W.: Personal interview by Suppanat Sakprasert dd. May 15, 2024 (2024)
17. Nithiyuwith, T., Treenuntharath, T.: The development of Thai artificial intelligence Chatbot for supporting academic consultancy for tertiary students. Life Sci. Environ. J. **21**(2), 453–467 (2020). https://ph01.tci-thaijo.org/index.php/psru/article/view/242312
18. Sakkampang, D., Mansourmahani, S.: The effects of the use AI Chatbot on communicative writing skill and self-regulated learning strategies of Thai university students. J. Man Soc. **9**(2), 119–138 (2023). https://so06.tci-thaijo.org/index.php/husocjournal/article/view/268899
19. Thongchai, P.: Role of language teacher in Thailand 4.0. J. MCU Humanit. Rev. **3**(1), 99–106 (2018). https://so03.tci-thaijo.org/index.php/human/article/view/138232
20. Thanaprakopkorn, T.: Factors related to the impact of artificial intelligence on financial service of Siam commercial bank: a case study of Bangkok metropolitan region. MUT J. Bus. Admin. **17**(1), 55–72 (2020)
21. Patthararangsarith, P.: The impact of digital disruption on agriculture of Thailand. J. Ind. Educ. **18**(2), 1–5 (2019). https://ph01.tci-thaijo.org/index.php/JIE/article/view/212336
22. Meesad, P.: Thai fake news detection based on information retrieval, natural language processing and machine learning. SN Comput. Sci. **2**(6), 425 (2021). https://doi.org/10.1007/s42 979-021-00775-6
23. Kedthawon, S.: A systematic literature review using artificial intelligence and fake news detection. Interdiscip. Acad. Res. J. **4**(1), 603–616 (2024). https://doi.org/10.60027/iarj.2024. 273480
24. Wongsupalak, N., Jirakunkanokm, P.: Development of factors and indicators of service quality affecting to customer satisfaction and loyalty in financial services industry: Thai text analytics with artificial intelligence. RMUTT Global Bus. Econ. Rev. **17**(2), 114–133 (2022). https:// so03.tci-thaijo.org/index.php/RMUTT-Gber/article/view/262214
25. Khunthong, C., Nanakorn, P.: Copyright protection of works created by artificial intelligence. Rajapark J. **17**(55), 17–31 (2023). https://so05.tci-thaijo.org/index.php/RJPJ/article/ view/266703
26. Butr-indr, P.: The law and artificial intelligence. J. Fac. Law **47**(3), 491–511 (2018). https:// so05.tci-thaijo.org/index.php/tulawjournal/article/view/195250
27. Ruangariyapuk, N.: A causal correlation model of the artificial intelligence acceptance towards marketing performance of Thai airlines. MUT J. Bus. Admin. **20**(2), 88–111 (2023)
28. Wasutharat, T.: Factors influencing acceptance of using financial services through AI Chatbot of commercial banks in central region of Thailand. J. KMITL Bus. Sch. **11**(1), 39–51 (2021). https://so02.tci-thaijo.org/index.php/fam/article/view/247735
29. Chumongkol, P.: A structural equation model of Thai digital citizens' artificial intelligence adoption in marketing communication. J. Public Relat. Advertising **13**(1), 94–113 (2020)
30. Chaisuwan, B.N., Chantamas, M.: การ จัด กลุ่ม บุคลากร วัย ทำงาน ตาม ระดับ การ ยอมรับ เ ทคโนโลยี ปัญญา ประดิษฐ์. J. Commun. Arts **42**(1), 99–117 (2024)

31. Hatani, F.: Artificial intelligence in Japan: policy, prospects, and obstacles in the automotive industry. In: Khare, A., Ishikura, H., Baber, W.W. (eds.) Transforming Japanese business. FBF, pp. 211–226. Springer, Singapore (2020). https://doi.org/10.1007/978-981-15-0327-6_15

32. Zhou, L.: A historical overview of artificial intelligence in China. Sci. Insights **42**(6), 969–973 (2023). https://doi.org/10.15354/si.23.re588

33. Naudé, W.: Artificial intelligence: neither Utopian nor apocalyptic impacts soon. Econ. Innov. New Technol. **30**(1), 1–23 (2021). https://doi.org/10.1080/10438599.2020.1839173

34. SEAMEO STEM-ED: Artificial intelligence and its comprehensive history. Southeast Asian Ministers of Education Organization (SEAMEO) (2023). https://seameo-stemed.org/blog/tec hnology-fact-artificial-intelligence-and-its-comprehensive-history/

35. Poovarawan, Y.: Personal interview by Suppanat Sakprasert dd. May 10, 2024 (2024)

36. Palasri, S., Huter, S., Wenzel, Z.: The history of the internet in Thailand. University of Oregon libraries, Oregon (2013 [1999]). https://nsrc.org/sites/default/files/archives/case-studies/TH_ history_2013.pdf

37. Humanoid Robotics Institute (n.n.): WABOT -WAseda roBOT-. Waseda University. https://www.humanoid.waseda.ac.jp/booklet/kato_2.html

38. Garvey, C.: Artificial intelligence and Japan's fifth generation: the information society, neoliberalism, and alternative modernities. Pac. Hist. Rev. **88**(4), 619–658 (2019). https://www.jstor.org/stable/26861060

39. Anyoha, R.: The history of artificial intelligence. science in the news, Harvard graduate school of the arts and sciences (2017). https://sitn.hms.harvard.edu/flash/2017/history-artificial-intell igence/

40. Russell, S., Norvig, P.: Artificial Intelligence: A Modern Approach, 4th edn. Pearson (2020)

41. Kawtrakul, A.: Personal interview by Suppanat Sakprasert dd. February 20, 2024 (2024)

42. Koanantakool, H.T.: The Internet in Thailand: the perpetual chronicles of internet events in Thailand. NECTEC (2001). https://nectec.or.th/users/htk/milestones.html

43. The Straits Times: The 1990s. Singapore Press Holdings (2015). https://graphics.straitsti mes.com/STI/STIMEDIA/Interactives/2015/10/35-years-of-ict/supercharging-singapore/ the-1990s.html

44. Jasanoff, S., Kim, S.H.: Containing the atom: sociotechnical imaginaries and nuclear power in the United States and South Korea. Minerva **47**, 119–146 (2009). https://doi.org/10.1007/ s11024-009-9124-4

45. Az Zahra, A., Nurmandi, A.: The strategy of develop artificial intelligence in Singapore, United States, and United Kingdom. IOP Conf. Ser. Earth Environ. Sci. **717**(1) (2021). https://doi.org/10.1088/1755-1315/717/1/012012

46. Thajchayapong, P., Changgom, G.: Supervising the Internet in Thailand. Media Asia **25**(2), 63–66 (1998). https://doi.org/10.1080/01296612.1998.11726549

47. Leightner, J.E.: Thailand's financial crisis: its causes, consequences, and implications. J. Econ. Issues **41**(1), 61–76 (2007). https://doi.org/10.1080/00213624.2007.11506995

48. Farrell, W.C., Phungsoonthorn, T.: Generation Z in Thailand. Int. J. Cross Cult. Manage. **20**(1), 25–51 (2020). https://doi.org/10.1177/1470595820904116

49. Magpanthong, C.: Thailand's evolving Internet policies: the search for a balance between national security and the right to information. Asian J. Commun. **23**(1), 1–16 (2013). https://doi.org/10.1080/01292986.2012.717089

50. Jasanoff, S., McGonigle, I., Stevens, H.: Science and technology for humanity: an STS view from Singapore. East Asian Sci. Technol. Soc. **15**(1), 68–78 (2021). https://doi.org/10.1080/ 18752160.2021.1877034

51. Leesa-Nguansuk, S.: Easyrice uses artificial intelligence for food revolution. Bangkok Post (2023). https://www.bangkokpost.com/business/general/2632183/easyrice-uses-artificial-int elligence-for-food-revolution

52. Sharma, K.: AI powered technology transforming Thailand hotels. Thaiger (2023). https://the thaiger.com/hot-news/technology/ai-powered-technology-transforming-thailand-hotels
53. McFaul, C., Chahal, H., Gelles, R., Konaev, M.: Assessing South Korea's AI ecosystem. Center for security and emerging technology (2023). https://cset.georgetown.edu/publication/assess ing-south-koreas-ai-ecosystem/
54. Keyvan, Ö.Z.: ABD - ÇİN REKABETİNDE TAYLAND'IN HEDGİNG STRATEJİSİ. *Pamukkale Üniversitesi Sosyal Bilimler Enstitüsü Dergisi* [Pamukkale University Journal of Social Sciences Institute] **52**, 313–331 (2022). https://doi.org/10.30794/pausbed.1079090
55. Baxter, W.: Thailand 4.0 and the future of work in the Kingdom. International Labour Organization (2017). https://www.ilo.org/resource/thailand-40-and-future-work-kingdom
56. Ministry of Digital Economy and Society: Digital Thailand – AI ethics guideline. Bangkok (2020). https://www.etda.or.th/getattachment/9d370f25-f37a-4b7c-b661-48d2d730651d/Dig ital-Thailand-AI-Ethics-Principle-and-Guideline.pdf.aspx?lang=th-TH
57. Chokesuwattanakul, P., Boonaramrung, P., Kowpatthahakij, P., Unpat, C., Thipsumritkul, T., Chartbunchachai, Y.: Thailand artificial intelligence guidelines 1.0. Center for Research on Law and Development, Bangkok (2022). https://www.law.chula.ac.th/wp-content/uploads/2023/03/TAIG-20230222.pdf
58. AI Ethics Committee NSTDA: AI ethics guidelines. National Science and Technology Development Agency (NSTDA), Bangkok (2022). https://bact.cc/f/2022/10/20220831-aw-book-ai-ethics-guideline.pdf
59. Suriyawongkul, A.: A collection of documents on Thai AI governance (Oct. 2022 / Mar. 2024) [Translated from รวมเอกสารข้อเสนอการกำกับกิจการ AI ของไทย (ต.ค. 2565 / มี.ค. 2567)]. bact' is a name (2022). https://bact.cc/2022/thailand-ai-regulations/
60. Kreps, D., de Roche, M., Gotterbarn, D., Havey, M.: IFIP code of ethics and professional conduct. International Federation for Information Processing (IFIP) (2021). https://www.ipt hree.org/ifip-code-of-ethics/
61. Hongladarom, S., Bandasak, J.: Non-western AI ethics guidelines: Implications for intercultural ethics of technology. AI Soc. (2023). https://doi.org/10.1007/s00146-023-016 65-6
62. Toupin, S.: Shaping feminist artificial intelligence. New Media Soc. **26**(1), 580–595 (2024). https://doi.org/10.1177/14614448221150776
63. Article 19: Thailand: Computer Crime Act (2017). https://www.article19.org/data/files/med ialibrary/38615/Analysis-Thailand-Computer-Crime-Act-31-Jan-17.pdf
64. National Electronics and Computer Technology Center (NECTEC): AI Thailand | Thailand national AI strategy and action plan (2022–2027) (2022). https://ai.in.th/en/about-ai-thailand/
65. Electronic Government Agency (EGA): (Draft) five-year Thailand digital government development plan (2017–2021). (2017). https://www.dga.or.th/wp-content/uploads/2023/07/Dig ital-Government-Development-Plan-2017-2021-Eng-Version.pdf
66. Pipatanakul, K, et al.: Typhoon: Thai large language models (2023). arXiv. https://doi.org/10.48550/arXiv.2312.13951
67. Smart Nation Singapore: National AI strategy. Smart nation and digital government office, Singapore (2024). https://www.smartnation.gov.sg/nais/
68. Zhang, H., Khanal, S.: To win the great AI race, China turns to Southeast Asia. Asia Policy **19**(1), 21–34 (2024). https://doi.org/10.1353/asp.2024.a918871
69. Pau, J., Baker, J., Houston, N.: Artificial intelligence in Asia: preparedness and resilience. Asia Business Council (2017). https://www.asiabusinesscouncil.org/docs/AI_briefing.pdf
70. Sheehan, M.: China's AI regulations and how they get made. Carnegie endowment for international peace (2023). https://carnegieendowment.org/research/2023/07/chinas-ai-regulations-and-how-they-get-made?lang=en

71. Japan External Trade Organization (JETRO): Japan's U.S. investment dynamic 2023 (2023). https://www.jetro.go.jp/ext_images/usa/2023/Japan-US-Investment-Report/Japans_US_Investment_Dynamics2023.pdf
72. Chandran, R.; Despite international hires, TikTok is Chinese at its core [Email newsletter dd. 7 May 2024]. Rest of World (2024)
73. United Nations ESCAP (n.n.): Technology and innovation policies. United Nations Economic and Social Commission for Asia and the Pacific (ESCAP). https://www.unescap.org/our-work/trade-investment-innovation/technology-innovation-policies/artificial-intelligence
74. Joyce, K., et al.: Toward a sociology of artificial intelligence: a call for research on inequalities and structural change. Socius 7 (2021). https://doi.org/10.1177/2378023121999581
75. Tan, J.-E.: Governance of artificial intelligence (AI) in Southeast Asia. EngageMedia (2021). https://engagemedia.org/wp-content/uploads/2023/06/Engage_Report-Governance-of-Artificial-Intelligence-AI-in-Southeast-Asia_06062023.pdf
76. United Nations ESCAP: Frontier technologies for sustainable development in Asia and the Pacific. United nations economic and social commission for Asia and the Pacific (ESCAP) (2018). https://www.unescap.org/publications/frontier-technologies-sustainable-development-asia-and-pacific
77. Alizadeh, F., Stevens, G., Esau, M.: I don't know, is AI also used in airbags? An empirical study of folk concepts and people's expectations of current and future artificial intelligence. I-com 20(1), 3–17 (2021). https://doi.org/10.1515/icom-2021-0009

Generative AI-Augmented Decision-Making for Business Information Systems

Endrit Kromidha[1]([⊠]) [iD] and Robert M. Davison[2] [iD]

[1] University of Birmingham, Birmingham, UK
e.kromidha@bham.ac.uk
[2] City University of Hong Kong, Hong Kong, Hong Kong
isrobert@cityu.edu.hk

Abstract. The integration of Generative Artificial Intelligence (GAI) in decision-making has ushered in a new era of opportunities and challenges for organizations. However, due to the way GAI algorithms work, the propagation of social biases and the lack of transparency on how they use data raises concerns about the autonomy and control of human decision-making powered by such systems. In this experimental research paper, we contrast the answers that ChatGPT, a popular GAI tool, can give us in the context of decision-making with what we know about such processes from the management and business information systems literature. Our findings suggest that GAI can facilitate the organization of information and options for making decisions. However, without a moral and ethical stance, the responsibility for the decisions remains with the human actor. Suggesting a collaborative approach between humans and GAI, we reflect on the changes to learning and adaptation patterns that need to happen on both sides to reform the way we make GAI-augmented decisions.

Keywords: Generative Artificial Intelligence (GAI) · ChatGPT · Decision-Making · Business Information Systems

1 Introduction

Artificial Intelligence (AI) tools constitute an emerging, developing yet already widely-available addition to the compendium of tools that we can employ in a variety of contexts and situations. By definition, AI uses machine learning, neural networks, and related technologies to generate new content based on patterns identified from training data [1]. Recent literature reviews on the topic identify a gap between AI research and practice [2], and limited knowledge on how AI can add business value [3]. In this context, a key domain where AI may prove valuable, but about which we know little about, is decision-making, an activity that has characterized the human experience since time immemorial.

Decision-making, at least where it is undertaken rationally and responsibly, is an activity that may incorporate a variety of subroutines that bring structure to unstructured

© IFIP International Federation for Information Processing 2024
Published by Springer Nature Switzerland AG 2024
R. M. Davison and D. Kreps (Eds.): HCC 2024, IFIP AICT 719, pp. 46–55, 2024.
https://doi.org/10.1007/978-3-031-67535-5_5

decision processes [4]. Although some decisions seem undeniably intuitive, instinctive or reflexive [5, 6], in most organizational contexts a decision maker needs to be familiar with potentially a large volume of information, and in making the decision can choose to be aware of the impact of the decision on multiple stakeholders. Accessing and digesting this information, even when using information technologies, as well as considering impacts, can consume considerable intellectual resources [7]. However, AI is likely to reduce the human resource requirements markedly, and thus may offer an attractive option to the harried decision maker. This is not to suggest that AI alone can or should make decisions, but simply to reveal the potential benefits that including an AI tool in the decision-making process may enable. In this research in progress paper, we explore the potential value that AI may bring to the decision maker in order to answer our research question: How can AI support rational decision-making undertaken by human actors?

Following this introduction, we briefly review the literature and explain our methods, which involved our interacting with the free version of ChatGPT 3.5, a popular AI. AI technologies have the capability to not only give us answers, but also to justify them. Furthermore, they can justify their suggestions, which may also help us to make informed decisions. Given this potential, in this study we ask ChatGPT to tell us how it can help with decision-making, and we contrast its answers with what we know from the business management and information systems (IS) literature. We explore the implications of the AI response in the light of current decision-making practices and the related literature, before concluding the paper with the research and practice implications that stem from this work.

2   Literature Review: Computer Assisted Decision-Making and AI

Decision-making has always been an inseparable feature of human civilization. An early decision-making theory informed by economic and psychological perspectives assumes that actors are rational when they make choices to maximize their utility [8]. Regardless of unrealistic assumptions drawing attention to normative implications, bounded rationality, power to win battles of choice, and chance are important in strategic decision-making [9]. Mintzberg et al. [4], in their seminal paper about the structure of unstructured strategic decisions, identified three sets of supporting routines and six sets of dynamic factors that pertain to the identification, development and selection decision-making phases. Structured approaches like this have been welcomed in IS research where technology increases efficiency and effectiveness by helping to process data for decision-making [10]. In this context, where humans need to work with technology to make decisions, adaptive approaches to decision-making, responsibilities [11] and cognition [12] have been at the center of research and practice debates.

A key problem in computer-assisted decision-making is the behavioral dilemma between decision quality and conservation of effort [7]. Despite their lack of intrinsic and conscious responsibility, information technologies are often considered as an aid for the responsible human actor. A good balance between intuitive and technocratic decision-making is needed [13], adjusted for industry type and dynamism [14]. However, while technologies might be amoral, technological choices used in decision-making have moral consequences [15, 16]. As we build technological solutions like GAI tools that are able to mimic human decision-making capabilities or to process information and propose solutions, it is also necessary to revisit how we use technology to make decisions.

Earlier AI models have been instrumental to decision-making with their predictive machine learning capabilities [17, 18], but new AI models are often defined as being generative because they can create new elaborated content and responses. Therefore, we refer to them as generative artificial intelligence models, or otherwise GAI in this study, with ChatGPT being a prime popular example at the moment. GAI can help organizations with process automation, cognitive insight and cognitive engagement [19], listed in order of research interest based on a recent literature review [20]. However, we must distinguish between the human and GAI capabilities to produce new knowledge, with the latter being more suitable for decentralization and deferred temporary binding [21]. Although GAI might be able to provide highly plausible and intelligent sounding responses, it might lack intelligence per se in imagining new futures [22]. At the same time, GAI's use in decision-making is constrained by its lack of culpability, moral accountability, and public accountability and its inability to exercise active responsibility [23]. How we can use these characteristics of GAI to maintain our position of power in decision-making is a fundamental operational and ethical debate that we engage with in this study.

3   Methodology, Data and Analysis

We employ a field experiment as our research methodology, since, even on digital platforms for business and entrepreneurship, research has typically involved human interaction [24]. Following similar principles, we adopt a new approach in this study where empirical evidence is generated by interacting with ChatGPT, the most well-known GAI tool to date (Table 1).

Table 1. Records of interactions on decision-making support with ChatGPT.

Nr	17 October 2023	20 December 2023	31 January 2024
1	Information and Research	Information Gathering	Information Gathering
2	Pros and Cons	Analysis	Exploring Options
3	Data Analysis	Pros and Cons	Decision Analysis,
4	Scenario Analysis	Scenario Planning	Decision-Making Frameworks
5	Decision Frameworks	Clarifying Goals	Clarifying Goals and Values
6	Recommendations	Brainstorming	Scenario Planning
7	Clarification	Decision Frameworks	Reflective Listening
8	Creative Problem-Solving	Clarifying Values	
9	Goal Alignment		
10	Emotional Considerations		

On the 17th of October 2023, the lead author based in Europe asked ChatGPT the following question: "How can you help me in decision-making?" ChatGPT's answer delineates 10 distinct areas as follows: 1. Information and Research; 2. Pros and Cons; 3. Data Analysis, 4. Scenario Analysis; 5. Decision Frameworks, 6. Recommendations; 7. Clarification; 8. Creative Problem-Solving; 9. Goal Alignment, and 10. Emotional Considerations.

On the 20th of December 2023, the second author based in Asia asked the same question to ChatGPT, receiving a similar but not the same answer, listing only eight areas: 1. Information Gathering, 2. Analysis, 3. Pros and Cons, 4. Scenario Planning, 5. Clarifying Goals, 6. Brainstorming, 7. Decision Frameworks, and 8. Clarifying Values.

On the 31st of January 2024, the first author asked the same question again to Chat-GPT, receiving an answer with only seven points as follows in this order: 1. Information Gathering, 2. Exploring Options, 3. Decision Analysis, 4. Decision-Making Frameworks, 5. Clarifying Goals and Values, 6. Scenario Planning, 7. Reflective Listening.

We discuss and recombine these areas and their explanations in the light of the extant IS and business literature in order to understand the potential of GAI to change decision-making in organizations. Practical implications include a critical examination of the ethical issues that could result from such changes in decision-making style.

Responses from ChatGPT were contrasted with secondary data from the summary report: 'AI ecosystem survey: informing the National AI Strategy' that includes GAI models, conducted by the Alan Turing Institute in collaboration with the AI Council in the UK in June 2021. The survey used asked over 390 business representatives to comment on the following areas of the AI Roadmap: Research and innovation, Skills, Data, infrastructure and public trust, National, cross-sector adoption. The focus of our analysis was on the skills and capabilities and their limitations for UK businesses to make use of the full potential of GAI.

4 Findings and Discussion

ChatGPT starts its first reply like this: *"I can assist you in decision-making by providing information, analysis, and suggestions based on the input you provide. Here's how I can help:"* It is important to note that while ChatGPT does accept the responsibility to offer an answer, it also leaves the responsibility to the user and the input provided. The disclaimer is elaborated at the end of answer, after the 10 areas are provided, reminding the reader that ChatGPT does not take any responsibility for its answers: *"It's important to note that while I can provide valuable assistance, the final decision should always be made by you, taking into account your personal judgment, values, and the specific context of the decision. Additionally, I may not have access to the most current information or specific details, so it's essential to cross-check information when making important decisions."* This scenario sets the scene to understand the following options and put them in perspective for computer assisted decision-making with GAI.

From the preliminary analysis, Mintzberg et al.'s [4] decision-making phases of identification, development and selection were considered to categorize the areas where ChatGPT and similar GAI tools can help with decision-making. Data and information inputs are not much different than before, but what is different and interesting is that all supporting routines and dynamic factors are now replaced by the black box of one single algorithm. After the analysis, the answers from ChatGPT lead us to suggest four decision-making phases and processes assisted and influenced by GAI: 1. Understanding the problem, 2. Evaluating options, 3. Choosing decision-making frameworks, and 4. Balancing emotions and alignment.

4.1 Understanding the Problem

Information and research are the first steps for making any decision. While ChatGPT does not know the most recent information and news, it can still be useful to navigate historic data it has access to. The role of GAI in this case is similar to that of a quasi-omniscient research assistant. Research with executives confirms that the quality of information and research is important for rational and well-informed decision-making [25]. In this regard, ChatGPT has limitations of data coverage that do not go beyond 2021, and we do not really know how the algorithm uses information from human actors, e.g., those that are supplied by the user. Facts, explanations, and background information are listed separately in its statement above, but in practice, it is difficult to distinguish them from each other. What we can do is challenge the answers a GAI tool gives using its own computing power.

To be sure, ChatGPT offers clarification help. In business venturing, clarification is used to secure buy-in when difficult collegial decisions have to be negotiated and made in an accountable manner [26]. Formal control can help towards decision-making clarity and innovation performance [27]. Similarly, a GAI tool can offer such assistance by allowing the decision-maker to delve into various options and outcomes, but clarifications will only be as good as the data inputs and engagement of users with them is. As such, GAI can offer clarifications through contextualization based on the breadth of its intelligence, but for specific cases, the user will have to seek detailed clarifications elsewhere.

4.2 Evaluating Options

Using its computing power, ChatGPT is not only a search box and memory machine, but also able to analyze and interpret data to make sense of it for decision-making. Any GAI tool is clearly built around data, usually a lot of it, but the technologies it uses are limited to computer vision, machine learning, natural language processing, robots and speech recognition [28]. Simple computational operations are possible with GAI, but for expert analytics, specialized software systems and expertise is needed.

To make things even simpler, pros and cons are a simple way of analyzing options offered by ChatGPT. In business, evaluating pros and cons is about trade-offs, risks and even swaps [29]. A powerful technique for making hard decisions by listing options and outcomes, evaluating pros and cons requires an unbiased approach that AI can offer because it does not have any vested interest in the answer. For the same reason, personal factors that can influence decisions are left again to the discretion of the decision-maker.

For more complex insights, scenario analysis is being offered by ChatGPT, a technique that has been considered "an 'art' that lacks theoretical and methodological rigor" [30]. As advancements have been made to address this use, the moral imperative that should guide scenario analysis remains the most important element. ChatGPT explains that it does not apply morals or ethical principles on its own, and responses can be influenced by the data it was trained on. As such, scenario analysis for developing decisions has value only in quite general terms.

Finaly, without being able to make decisions for us, ChatGPT can offer recommendations which are powerful contributors that influence decisions, as we can see when products are suggested to us online [31]. In management, evaluators need to assess the value of their recommendations based on the use context, evaluator's role, and the typology recommendations [32]. This, however, is not possible with GAI, because it is not possible to assess the level of its expertise as an evaluator making recommendations. All we can do is trust that the vast amount of information it draws on to offer recommendations will be sufficient to generate a broadly relevant answer for us from which we can further customize our actual decisions.

4.3   Choosing Decision-Making Frameworks

ChatGPT offers to explain, and possibly apply decision-making frameworks and methodologies like SWOT analysis, cost-benefit analysis, or decision trees, which can guide your decision-making process. Leaders, consciously or unconsciously, use frameworks to make decisions, avoid disorder and address simple, chaotic, complicated or complex problems [33]. The tools listed by ChatGPT above are well-known frameworks used in business management, and while they work, they also show the tendency of a GAI system built with public big data to remain generally trivial by focusing on what the majority would expect, rather than what a specific decision might require. Frameworks beyond the common knowledge could be accessible through GAI tools, but a higher potential of error margin needs to be taken into consideration.

Creative problem solving through in brainstorming or generating innovative solutions similar to the approach used in understanding the problem, but it can serve also to reflect on the process of how we make decisions. Creative problem solving is a challenge because under pressure, managers prefer tried and tested options instead of considering new things [34]. A GAI tool could be a powerful tool in this case because it could extract, process, and combine many creative alternatives to a problem that requires a decision. This is probably the most powerful feature of GAI for assisting decision-making because it combines its ability to navigate through options the decision-maker might not have thought about very quickly and efficiently. This is only influenced by the data it has been trained on, and its users which is one of its main limitations as ChatGPT was trained on datasets like Wikipedia, BookCorpus for books, and Common Crawl for web content which are mainly in English [35], potentially restricted from other cultures and languages.

4.4   Balancing Emotions and Alignment

ChatGPT acknowledges that sometimes, decisions are influenced by emotions, offering to provide an objective perspective to consider emotional factors in a rational way. Covering the relationship between emotions and decision-making is a broad topic, but research confirms that GAI can now demonstrate cognitive capabilities, such as sensing emotions, that were previously not possible [36]. While for humans we have commonly linked emotions to intuition in managerial decision-making [37], in the case of GAI we can only argue that the way it can sense emotions is different. What we humans and GAI have in common is the ability to adapt and learn through interactions. As GAI becomes better at feeling emotions as we do [38], and as we engage more with such tools, it is inevitable that we will learn to understand each other better on an emotional as well as cognitive level that together could lead to better decision-making.

ChatGPT's offer of rationality through assistance for the alignment of options with goals, values and priorities is another way of showing emotional support. With increasing complexity in business information systems and organizations, multiple goals emerge and need to be aligned at the same time. This creates a burden for decision-makers. Without losing its ability to offer more choices and options, GAI in this case tries to apply the reverse logic of consolidation for decision-making by identifying conceptual relationships. In business and IS research, goal alignment is strongly linked to process alignment and strategic alignment [39] guided by efficiency, agility, and diversification principles [40]. The detachment of GAI from the user's personal and emotional characteristics could offer a valuable perspective that is informed by the data and their annotations that GAI uses, in addition to the questions being asked. This could help decision-making by providing logical links and context to see the forest when we risk missing it by focusing on the trees.

4.5   Contrasting GAI Answers with Human Skills and Capabilities

From the analysis of the AI Council survey, most respondents agree to have the necessary skills and knowledge to understand for themselves when value could be gained from utilizing AI (125/390), and they confirm that there are plenty of opportunities to train themselves about it (112/390). Respondents strongly disagree that there are significant barriers to recruiting and retaining top AI talent within the UK (132/390). These answers suggest that GAI is perceived as a tool that has potential to help their organizations with the right human agency being put in control for its use and decision-making. To better understand these results, it is important to put in perspective that over 30% of respondents work in academia, and over 40% work in the industry with very few from the public sector, learned society, professional bodies or the third sector. Over 60% of them being male, based in London (Over 40%) and working as directors (over 35%), it is not surprising that for professional decision-making, ChatGPT still offers limited solutions.

5   Research and Practice Implications

It is clear from this research that GAI-augmented decision-making processes are not very different from human-like decision-making in organizations. The structure of seemingly unstructured decision-making aids follows the familiar path of identification, development and selection [4]. However, what GAI does differently is replacing routines and dynamic factors that are essential for human decision-making processes with a single algorithm that is able to quickly and efficiently process data and human inputs to generate results. This requires a reformulation of the computer-assisted decision-making dilemma between decision quality, or conservation of effort [7]. Considering the rich toolkit of aids GAI offers to make decisions, combining them together could result in very efficient decision-making that requires less effort than any other method. As a result, the effort problem seems to have become irrelevant with GAI.

What emerges is a new set of problems related to lack of culpability, moral account-ability, public accountability and the ability to exercise active responsibility [23] as evidenced in this study on decision-making. Morally, the human actor using GAI, as with every other technology, will have to accept the responsibility that comes with computer-assisted decisions. The reformulated dilemma in GAI-augmented decision-making now lies between decision quality and decision amorality. That does not mean that GAI's morals are negative, but simply that it does not have any particular guiding morals and ethical norms except those that could be embedded in it from the data inputs and our interactions with it. In this sense, GAI tools are digital representations of collective morality or amorality, which in a postmodernist society could be fairly positionless.

If the training data included materials from societies that do not subscribe to the post-modernist tradition, e.g., those where theological values are important, then these could alter that morality. Such an effect could also be realized if we had initially asked a question along the lines of "How can you help me in decision-making bearing in mind the x requirements of y?" where x could refer to specific moral codes and y could refer to specific theological institutions. This suggests that people who rely on GAI need to be more rather than less precise in how they interact with it, providing enough contextual detail to ensure that the answers are contextually appropriate.

Finally, no other technologies before have been able to justify their reason for being, and for being used, better than GAI tools. However, taking a clear ethical stance, or assuming particular moral standards in the absence of any specific instructions to do so, is in itself a decision that GAI is designed neither to take for either itself or us because this would imply related consequences and responsibilities. Considering them as simple information technology aids means disregarding their full intelligence capabilities. The risks associated with moral and ethical decisions may be considered unpredictable and therefore unquantifiable in the form of insurance for damages caused by GAI-augmented decision-making systems. However, as we use GAI more, learn about it, and it learns about us, we should have a better understanding of the risks involved, and devise strategies to mitigate them efficiently. Until then, in a gamified exploratory experience with GAI, we can only use its capabilities to improve our decision-making. This, for now, means learning and perfecting new GAI-augmented decision-making routines and processes where computing efficiency is combined with our moral responsibility and accountability, and in doing so be able to intelligently evolve where GAI cannot.

References

1. Ooi, K.-B., et al.: The potential of Generative Artificial Intelligence across disciplines: perspectives and future directions. J. Comput. Inform. Syst. 1–32 (2023). https://doi.org/10.1080/08874417.2023.2261010
2. Nguyen, Q.N., Sidorova, A., Torres, R.: Artificial intelligence in business: a literature review and research agenda. Commun. Assoc. Inf. Syst. **50**(1), 175–207 (2022)

3. Enholm, I.M., Papagiannidis, E., Mikalef, P., Krogstie, J.: Artificial intelligence and business value: a literature review. Inf. Syst. Front. **24**(5), 1709–1734 (2022)
4. Mintzberg, H., Raisinghani, D., Theoret, A.: The structure of "unstructured" decision processes. Adm. Sci. Q. **21**(2), 246–275 (1976)
5. Järvilehto, L.: The Nature and Function of Intuitive Thought and Decision Making. Springer, Cham (2015).https://doi.org/10.1007/978-3-319-18176-9
6. Matzler, K., Bailom, F., Mooradian, T.A.: Intuitive decision making. MIT Sloan Manag. Rev. **49**(1), 12–16 (2007)
7. Todd, P., Benbasat, I.: The use of information in decision making: an experimental investigation of the impact of computer-based decision aids. MIS Q. **16**(3), 373–393 (1992)
8. Edwards, W.: The theory of decision making. Psychol. Bull. **51**(4), 380–417 (1954)
9. Eisenhardt, K.M., Zbaracki, M.J.: Strategic decision making. Strateg. Manag. J. **13**(S2), 17–37 (1992)
10. Molloy, S., Schwenk, C.R.: The effects of information technology on strategic decision making. J. Manage. Stud. **32**(3), 283–311 (1995)
11. Chu, Y.-Y., Rouse, W.B.: Adaptive allocation of decision making responsibility between human and computer in multitask situations. IEEE Trans. Syst. Manage. Cybern. **9**(12), 769–778 (1979)
12. Payne, S.J., Howes, A., Reader, W.R.: Adaptively distributing cognition: a decision-making perspective on human-computer interaction. Behav. Inform. Technol. **20**(5), 339–346 (2001)
13. Covin, J.G., Slevin, D.P., Heeley, M.B.: Strategic decision making in an intuitive vs. technocratic mode: structural and environmental considerations. J. Bus. Res. **52**(1), 51–67 (2001)
14. Bullini Orlandi, L., Pierce, P.: Analysis or intuition? Reframing the decision-making styles debate in technological settings. Manag. Decis. **58**(1), 129–145 (2020)
15. De George, R.T.: The Ethics of Information Technology and Business. Wiley, New Jersey (2008)
16. Johnson, D.G.: Technology with no human responsibility? J. Bus. Ethics **127**(4), 707–715 (2015)
17. Baird, A., Maruping, L.M.: The next generation of research on is use: a theoretical framework of delegation to and from agentic is artifacts. MIS Q. **45**(1), 315–341 (2021)
18. Agrawal, A., Gans, J., Goldfarb, A.: Prediction Machines, Updated and Expanded: The Simple Economics of Artificial Intelligence. Harvard Business Press, Harvard (2022)
19. Davenport, T.H., Ronanki, R.: Artificial intelligence for the real world. Harv. Bus. Rev. **96**(1), 108–116 (2018)
20. Collins, C., Dennehy, D., Conboy, K., Mikalef, P.: Artificial intelligence in information systems research: a systematic literature review and research agenda. Int. J. Inf. Manage. **60**, 102383 (2021)
21. Yoo, Y.: Evolving epistemic infrastructure: the role of scientific journals in the age of generative AI. J. Assoc. Inf. Syst. **25**(1), 137–144 (2024)
22. Russell, S.: Human Compatible: AI and the Problem of Control. Penguin, London (2019)
23. Santoni de Sio, F., Mecacci, G.: Four responsibility gaps with artificial intelligence: why they matter and how to address them. Philos. Technol. **34**, 1057–1084 (2021)
24. Lall, S.A., Chen, L.-W., Mason, D.P.: Digital platforms and entrepreneurial support: a field experiment in online mentoring. Small Bus. Econ. **61**(2), 631–654 (2023)
25. Citroen, C.L.: The role of information in strategic decision-making. Int. J. Inf. Manage. **31**(6), 493–501 (2011)
26. Gilmore, T.N., Kazanjian, R.K.: Clarifying decision making in high-growth ventures: the use of responsibility charting. J. Bus. Ventur. **4**(1), 69–83 (1989)

27. Schultz, C., Salomo, S., de Brentani, U., Kleinschmidt, E.J.: How formal control influences decision-making clarity and innovation performance. J. Prod. Innov. Manag. **30**(3), 430–447 (2013)
28. Rahman, M.A.: Five Core Technologies of Artificial Intelligence, Medium (2020). https://medium.com/born-to-lead/five-core-technologies-of-artificial-intelligence-5cea219f49ed
29. Hammond, J.S., Keeney, R.L., Raiffa, H.: Even swaps: a rational method for making trade-offs. Harv. Bus. Rev. **76**, 137–150 (1998)
30. Wright, G., Cairns, G., O'Brien, F.A., Goodwin, P.: Scenario analysis to support decision making in addressing wicked problems: pitfalls and potential. Eur. J. Oper. Res. **278**(1), 3–19 (2019)
31. Zhang, H., Zhao, L., Gupta, S.: The role of online product recommendations on customer decision making and loyalty in social shopping communities. Int. J. Inf. Manage. **38**(1), 150–166 (2018)
32. Iriti, J.E., Bickel, W.E., Nelson, C.A.: Using recommendations in evaluation: a decision-making framework for evaluators. Am. J. Eval. **26**(4), 464–479 (2005)
33. Snowden, D.J., Boone, M.E.: A leader's framework for decision making. Harv. Bus. Rev. **85**, 68–76 (2007)
34. Proctor, T.: Creative Problem Solving for Managers: Developing Skills for Decision Making and Innovation. Routledge, New York, USA (2010)
35. Thompson, A.D.: What's in my AI - A Comprehensive Analysis of Datasets Used to Train GPT-1, GPT-2, GPT-3, GPT-NeoX-20B, Megatron-11B, MT-NLG, and Gopher (2022). https://lifearchitect.ai/whats-in-my-ai/
36. Duan, Y., Edwards, J.S., Dwivedi, Y.K.: Artificial intelligence for decision making in the era of Big Data–evolution, challenges and research agenda. Int. J. Inf. Manage. **48**, 63–71 (2019)
37. Simon, H.A.: Making management decisions: the role of intuition and emotion. Acad. Manag. Perspect. **1**(1), 57–64 (1987)
38. Shank, D.B., Graves, C., Gott, A., Gamez, P., Rodriguez, S.: Feeling our way to machine minds: People's emotions when perceiving mind in artificial intelligence. Comput. Hum. Behav. **98**, 256–266 (2019)
39. Mehta, M., Hirschheim, R.: Strategic alignment in mergers and acquisitions: theorizing IS integration decision making. J. Assoc. Inf. Syst. **8**(3), 143–174 (2007)
40. Hoffmann, D., Ahlemann, F., Reining, S.: Reconciling alignment, efficiency, and agility in IT project portfolio management: Recommendations based on a revelatory case study. Int. J. Project Manage. **38**(2), 124–136 (2020)

Consideration of App Store-Related Cases and New Legislations in the EU, the US and Japan

Mika Nakashima[✉] [ID]

Faculty of Global Informatics, Chuo University, Tokyo, Japan
nakashima.77h@g.chuo-u.ac.jp

Abstract. Currently, investigations, lawsuits, and the introduction of new legislation, including new legislative proposals for ex-ante regulations concerning app stores on digital platforms, are unfolding across various jurisdictions worldwide. In the EU, the European Commission investigated Apple's app store. The Digital Markets Act—a comprehensive ex-ante regulation for digital platforms—came into effect on May 2, 2023, encompassing regulations specific to app stores. In the US, Epic Games filed a lawsuit against Apple and Google, and a bill known as the Open App Markets Act, targeting app stores, was proposed on June 11, 2021. In Japan, the Fair Trade Commission conducted an inspection of Apple, and Apple and Google were designated as app store-operating companies under the Act on Improving Transparency and Fairness of Digital Platforms on April 1, 2021. Moreover, on June 16, 2023, the Digital Market Competition Council, established under the Cabinet, publicly released the Competition Assessment of the Mobile Ecosystem Final Report, signaling a shift toward the legislation of ex-ante regulations for app stores. This paper explores app store-related cases and new legislation (or proposed legislation) in Japan, the US, and Europe, with the aim of examining the interpretation of Japan's Antimonopoly Act and identifying optimal approaches to new legislation.

Keywords: App Store · Mobile Ecosystem · Digital Platform · DMA · Open App Markets Act

1 App Store-Related Cases and New Legislations in the EU, the US and Japan

1.1 Europe

(1) Case

In the EU, on June 16, 2020, the European Commission initiated formal antitrust investigations to assess whether Apple's rules for app developers, governing the distribution of apps via the App Store, violate EU competition rules. These investigations specifically focus on the mandatory use of Apple's proprietary in-app purchase system and restrictions imposed on developers regarding their ability to inform iPhone and iPad users

R. M. Davison and D. Kreps (Eds.): HCC 2024, IFIP AICT 719, pp. 56–68, 2024.
https://doi.org/10.1007/978-3-031-67535-5_6

about alternative, more cost-effective purchasing options outside of apps. The investigations address the application of these rules to all apps that compete with Apple's own apps and services in the European Economic Area. They have resulted from separate complaints lodged by Spotify and an e-book/audiobook distributor, highlighting the impact of the App Store rules on competition in the realms of music streaming and e-books/audiobooks[1].

On February 28, 2023, the European Commission issued a Statement of Objections to Apple, outlining concerns about App Store rules for music streaming providers. The Commission is troubled by Apple's anti-steering obligations imposed on music streaming app developers, asserting that these obligations hinder developers from informing consumers about more affordable subscription options. The Commission preliminarily views Apple's anti-steering obligations as unfair trading conditions, potentially violating Article 102 of the Treaty on the Functioning of the European Union (TFEU)[2].

(2) New Legislation

In the EU, the Digital Markets Act (DMA), a comprehensive ex-ante regulation of digital platforms, came into effect on May 2, 2023, encompassing regulations specifically addressing app stores.[3]

Article 5 Obligations for gatekeepers

...

7. The gatekeeper is prohibited from mandating end users or business users to use, offer, or interoperate with, an identification service, a web browser engine or a payment service, or technical services that support the provision of payment services, such as payment systems for in-app purchases, of that gatekeeper in the context of services provided by the business users using that gatekeeper's core platform services.

Article 6 Obligations for gatekeepers susceptible of being further specified under Article 8

...

4. The gatekeeper shall allow and technically facilitate the installation and effective use of third-party software applications or software application stores that use or interoperate with its operating system. Additionally, it must allow these software applications or software application stores to be accessed through means other than

---

[1] European Commission (2020), Press release, Antitrust: Commission opens investigations into Apple's App Store rules, June 16, 2020.

[2] European Commission (2023), Press release, Antitrust: Commission sends Statement of Objections to Apple clarifying concerns over App Store rules for music streaming providers, 28 February 2023.

[3] REGULATION (EU) 2022/1925 OF THE EUROPEAN PARLIAMENT AND OF THE COUNCIL of September 14, 2022, on contestable and fair markets in the digital sector and amending Directives (EU) 2019/1937 and (EU) 2020/1828 (Digital Markets Act).

Apart from this, there is Competition and Markets Authority : CMA (2022), the Mobile ecosystems Market study final report, June 10, 2022, as a policy document in the UK.

the relevant core platform services provided by the gatekeeper. The gatekeeper, where applicable, is prohibited from hindering downloaded third-party software applications or software application stores from prompting end users to decide whether to set the downloaded software application or software application store as their default. The gatekeeper is required to technically enable end users who opt to set the downloaded software application or software application store as their default to easily implement this change.

The gatekeeper shall not be prohibited from taking, to the extent that they are strictly necessary and proportionate, measures to ensure that third-party software applications or software application stores do not compromise the integrity of the hardware or operating system provided by the gatekeeper. However, such measures must be duly justified by the gatekeeper.

Furthermore, the gatekeeper shall not be restricted from implementing, to the extent that they are strictly necessary and proportionate, measures and settings—other than default settings—that enable end users to effectively enhance security concerning third-party software applications or software application stores. These measures and non-default settings must be duly justified by the gatekeeper.

Thus, the DMA establishes that 1) mandatory payment/billing systems are prohibited, 2) outlinks are allowed, and 3) sideloading is allowed. It also stipulates that security protection on the part of digital platforms can be justified, albeit in a limited capacity (underlining by the author).

1.2   The United States

(1) Case

In the US, Epic Games, Inc., filed a lawsuit against Apple Inc. And Google LLC. Finding Apple's 30% App Store fee too high, Epic Games introduced its own payment system in Fortnite, a game operated by Epic Games. Apple claimed a violation of its terms and subsequently removed Fortnite from the App Store. Epic Games argued that this action violated Article 1 of the Sherman Act (a federal law) and California's Unfair Competition Law (UCL), bringing 10 counts. On September 10, 2021, the US District Court for the Northern District of California dismissed 9 counts, including the dispute of Article 1 of the Sherman Act, only accepting the UCL dispute.[4] Dissatisfied, Epic Games and Apple filed appeals, and on April 23, 2023, the United States Court of Appeals for the Ninth Circuit largely upheld the district court's decision.[5] Dissatisfied, Epic Games and Apple filed final appeals, and on January 16, 2024, Supreme Court of the United States denied final appeals.

Here, let us delve into the district court decision, which garnered attention as the inaugural judicial decision regarding the app store of a digital platform. Concerning the dispute of Article 1 of the Sherman Act, the district court scrutinized the app store's nature

---

[4] Epic Games Inc. v. Apple Inc. (2021), Case No. 4:20-cv-05640-YGR, N.D. Cal., Sept 10, 2021.

[5] Epic Games Inc. v. Apple Inc. (2023), Case No. 4:20-cv-05640-YGR, United States Court of Appeals for the ninth circuit, April 23, 2023.

as a two-sided market, relying on a three-step test (similar to the one used in the Amex Case Supreme Court Decision, where the two-sided market nature of credit cards was at issue). In the first step of this test, the plaintiff provides evidence of an anticompetitive effect related to the defendant's actions. In the second step, the defendant issues a rebuttal, asserting that their actions have a procompetitive justified reason. In the third step, the plaintiff must prove that such a reason could have been achieved through other means that are not anticompetitive. In the second stage of this case, Apple argued and successfully demonstrated that the restrictions on app distribution were for security reasons. In the third stage, Epic Games was unable to convincingly refute this, resulting in the rejection of the count. However, the district court did find the anti-steering provision's restriction of price-related information by Apple to be in violation of the UCL and ordered an injunction to the anti-steering provision.

Epic Games made a similar claim against Google, arguing that the 30% fee imposed by "Google Play" was excessive. In response, Epic Games introduced its own payment system in Fortnite. Google contended that this action violated its terms, leading to the removal of Fortnite from the app store. Epic Games asserted that this constituted a breach of Article 2 of the Sherman Act. On December 11, 2023, in the US District Court for the Northern District of California, the jury sided with Epic Games, supporting the assertion that, under Article 2 of the Sherman Act, Google was either acquiring or maintaining a monopoly.[6]

(2) New Legislation

In the US, multiple bills for digital platform regulations were proposed on June 11, 2021, including the bill for the Open App Markets Act to regulate app stores.[7]

SEC. 3. PROTECTING A COMPETITIVE APP MARKET

(a) Exclusivity and Tying: A covered company shall not

(1) require developers to use or enable an in-app payment system owned or controlled by the covered company or any of its business partners as a condition of the distribution of an app on an app store or accessible on an operating system;
(2) mandate, as a term of distribution on an app store, that pricing terms or conditions of sale be equal to or more favorable on its app store than the terms or conditions under another app store; or
(3) take punitive action or otherwise impose less favorable terms and conditions against a developer for using or offering different pricing terms or conditions of sale through another in-app payment system or on another app store.

(b) Interference with Legitimate Business Communications: A covered company shall not impose restrictions on communications of developers with the users of an app of the developer through the app or direct outreach to a user concerning legitimate business offers, such as pricing terms and product or service offerings.

---

[6] Epic Games Inc. v. Goole LLC. (2023), Case No. 3:20-cv-05671-JD, N.D. Cal., 11 Dec 2023.
[7] Open App Markets Act, S.270, Reported to Senate version, February 17, 2022.
Apart from this, there is Department of Commerce (2023), COMPETITION IN THE MOBILE APPLICATION ECOSYSTEM, February 2023, as a policy document in the US.

Nothing in this subsection shall prohibit a covered company from providing a user the option to offer consent prior to the collection and sharing of the user's data by an app.

(c) Nonpublic Business Information: A covered company shall not use nonpublic business information derived from a third-party app for competing with that app.

(d) Interoperability: A covered company that controls the operating system or operating system configuration on which its app store operates shall allow and provide readily accessible means for users of that operating system to

(1) choose third-party apps or app stores as defaults for categories appropriate to the app or app store;
(2) install third-party apps or app stores through means other than its app store; and
(3) hide or delete apps or app stores provided or preinstalled by the app store owner or any of its business partners.

Thus, the bill for the Open App Markets Act establishes that 1) mandatory payment/billing systems are prohibited, 2) outlinks are allowed, and 3) sideloading is allowed (underlining by the author).

1.3 Japan

(1) Case

Pursuant to the Act on Prohibition of Private Monopolization and Maintenance of Fair Trade: the Antimonopoly Act's Article 3 (Private Monopolization) and Article 19 (Unfair Transaction Practices, Paragraph 12: Trading on Restrictive Terms), the Fair Trade Commission in Japan conducted an inspection on the operation of the App Store, where Apple displays apps for the iPhone. The concern was that it restricted the business activities of companies providing reader apps in relation to the sales of digital content such as music, e-books, and videos. In response, Apple proposed improvement measures, including revising the stipulations of guidelines and allowing outlinks for reader apps. After considering this proposal, the Fair Trade Commission acknowledged that its concerns had been eliminated. Subsequently, on September 2, 2021, the Fair Trade Commission publicly announced that it had discontinued its inspection, confirming that Apple had implemented the improvement measures.[8]

(2) New Legislation

In Japan, the Act on Improving Transparency and Fairness of Digital Platforms (Transparency Act) was enforced on February 1, 2021, as a new legislation. Based on this law,

---

[8] Fair Trade Commission (2021), Regarding the Processes of the Suspected Antimonopoly Act Violation Case Regarding Apple Inc., September 2, 2021.

   Regarding this case, Shuya Hayashi (2022), The Antimonopoly Act and the Banning of Outlinks by Apple Inc., TKC Law Library, New Judicial Precedent Commentary watch ◆Economic Law No. 80 Document Number z18817009-00-120802110, Kazuo Tosa (2023) , Evaluation of Mandatory In-App Billing, the Banning of Outlinks, Etc., in Terms of the Antimonopoly Act, NBL No. 1240, p. 90.

Apple Inc. And Google LLC were specified as app store-operating companies on April 1, 2021. The law designates companies providing digital platforms with a strong need for transparency and fairness in transactions, as "specific digital platform providers," subjecting them to specific regulations. This law is premised on independent and active efforts on part of the digital platform providers to improve transparency and fairness, with governmental involvement and regulation limited to the necessary minimum. It establishes a general framework of regulations through law, using the *co-regulation* method, entrusting specific details to the independent efforts of companies. The law mandates specific digital platform providers to disclose information such as transaction conditions, independently maintain procedures and systems, and submit an annual report with a self-evaluation of implemented measures and business summaries.[9] Following the enforcement of this law, the Ministry of Economy, Trade, and Industry implemented the first "monitoring review" on December 22, 2022. The review contained the following evaluations: "It is hoped that Apple Inc. And Google LLC will continuously work toward mutual understanding with user companies, such as giving detailed explanations on the relationship between costs and fees involved with app store operation and on the ideal state of cost burdens, and engaging in discussions with user company groups, etc.," and "Moreover, in regard to changes in rules related to payment measures, *it is important that they be actually used by user companies*, so future trends will be closely observed with the inclusion of evaluations by user companies" (italics by author).[10] On December 5, 2023, the second monitoring review was made public. It acknowledged Apple's implementation of the external link account entitlement request (that allows outlinks on reader apps) and Google's implementation of a pilot program for user choice billing (UCB). However, in actuality, Apple rejects "external link account entitlement requests" citing non-fulfillment of reader app requirements. Google accepts registrations for all qualified developers (except for game apps) in their implementation of the UCB pilot program. Therefore, the committee members raised concerns about the exclusion of game apps and noted limited progress in the usage of outlinks and UCB on reader apps, requesting continued improvements to ensure that they become substantial options.[11]

Furthermore, the Digital Market Competition Council set up under the Cabinet publicly announced the Competition Assessment of the Mobile Ecosystem Final Report on June 16, 2023, signaling a shift toward a new legislation of ex-ante regulations for app stores in addition to existing operations based on the Transparency Act.

---

[9] Ministry of Economy, Trade and Industry (2020), Points on the Act on Improving Transparency and Fairness on Specific Digital Platforms (Established on May 27, 2020, Distributed on June 3, 2020).

[10] Ministry of Economy, Trade and Industry (2022), Evaluation of the Transparency and Fairness of Specific Digital Platforms (General Item-Selling Online Mall and App Store Fields), December 22, 2022.

Regarding the italics, it is speculated that the reason for this wording is that despite the individual case where the Fair Trade Commission carried out an inspection and Apple revised the agreement related to payment measures as a way of accepting outlinks of reader apps, the reality was that companies providing reader apps were somehow not implementing outlinks.

[11] Ministry of Economy, Trade and Industry (2023), 2023 Fiscal Year Monitoring Meeting Related to the Transparency and Fairness of Digital Platforms, Solicitation of Opinions [General Item-Selling Online Mall/App Store Fields], December 5, 2023.

As a general issues, the report concludes that in Japan, mobile OS is in an oligopoly by Google and Apple. A different approach to that in the existing Antimonopoly Act is necessary, given the difficulties with regard to defining the market and recognizing illegality in relation to business models of digital platforms. In concrete terms, each issue will be met with a mix of two policies: a framework of ex-ante regulation and a framework of co-regulation. The target of the regulations should be companies that provide mobile OS of a certain size or larger and those that provide services of a certain size or larger in the field of app stores, browsers, and search engines. Companies subject to the regulations will submit reports to the enforcement agency on their state of compliance with the new regulations, and the said reports will be made public. Other measures to ensure effectiveness mentioned in the report include correction orders for violations, emergency stop orders (provisional measures), as well as injunction requests and compensation for civil damages and financial disadvantages (which uses the Antimonopoly Act's surcharge system as a reference).[12]

Specifically, the report covers various points, including specification changes in OS and browsers, matters related to app stores, restrictions on browser functionality, issues related to pre-installation and default settings, as well as the acquisition and use of data and access to functions of the OS, among others. With regard to app store-related matters, the report mentions items such as 1) the mandatory use of payment/billing systems, 2) restrictions on information dissemination and steering of other billing systems within the app, and 3) establishment of a competitive environment among reliable app stores (the acceptance of alternative app distribution channels).[13]

With regard to 1) the mandatory use of payment/billing systems, at present, third parties that use the App Store and the Google Play Store for selling in-app content are obliged to use the payment/billing system of Apple and Google. In addition, such parties are required to pay a certain fee through the said system (30% or 15%, etc.). In response to these issues, the report indicated certain directions. The Japanese government prohibits companies that provide app stores of a certain size or larger from mandating the use of their own payment/billing systems by developers. It makes those companies mandatory to establish fair, reasonable, and non-discriminatory usage conditions, including fees, for app store business users. Additionally, it addresses concerns about impediments to communication regarding refunds and other related matters through ongoing monitoring based on the Transparency Act.[14]

With regard to 2) restrictions on information dissemination and steering of other billing systems within the app, Apple and Google currently restrict developers using the App Store and the Google Play Store from providing information and steering within

---

[12] Secretariat of the Headquarters for Digital Market Competition Cabinet Secretariat (2023a), Competition Assessment of the Mobile Ecosystem Final Report, June 16, 2023.

   For a commentary of said report written by the Assistant Director to the Counsellor of the Secretariat of the Headquarters for Digital Market Competition Cabinet Secretariat, see Ken Tagane (2023), Regarding the Summary of the Competition Assessment of the Mobile Ecosystem Final Report, Fair Trade No. 875, p. 63.

[13] Supra note 12, Secretariat of the Headquarters for Digital Market Competition Cabinet Secretariat (2023a).

[14] Supra note 12, Secretariat of the Headquarters for Digital Market Competition Cabinet Secretariat (2023a).

apps. In concrete terms, steering users toward transactions outside the app store, for example, through the use of language within the app that encourages users to purchase digital goods outside the app or posting links within the app (outlinks), is prohibited. In response to this, The Japanese government requires companies providing app stores of a certain size or larger to allow developers, at no cost, to provide information or offer transactions, including different purchase conditions (such as outlinks and other in-app actions), to users acquired on the app store.[15]

With regard to 3) the establishment of a competitive environment among reliable app stores (sideloading, allowing alternative app distribution channels), at present, the installation of apps that do not originate from the App Store is in principle not allowed on iPhones. In response to this, the Japanese Government obliges companies providing OS of a certain size or larger to allow the effective use of alternative app distribution channels that ensure security and privacy. In concrete terms, the report considers allowing app distribution through alternative app stores pre-installed on the iPhone, through alternative app stores downloaded through the App Store with a prerequisite inspection by Apple, and through alternative app stores downloaded using a browser. However, allowing the direct download of an app from a website is not mandatory.[16]

2   Debates Surrounding the Competition Assessment of the Mobile Ecosystem Final Report by the Cabinet's Digital Market Competition Council

The Competition Assessment of the Mobile Ecosystem Final Report of the Cabinet's Digital Market Competition Council has gathered attention, and academics will likely analyze it in the future. However, as of now, only a few studies have been published, with one paper by an academic and another by an attorney. With regard to other information available currently, articles featuring interviews with academics engaged in the Digital Market Competition Council, opinions of those in charge of the Secretariat of the Headquarters for the Digital Market Competition Cabinet Secretariat, as well as public comments on the report, have been referenced. The public comments extended to 381 pages and included a diverse range of opinions from university professors, attorneys, companies, associations of companies, and consumer groups.

---

[15] Supra note 12, Secretariat of the Headquarters for Digital Market Competition Cabinet Secretariat (2023a).

In South Korea, the Telecommunications Business Act was amended on September 14, 2021, prohibiting the enforcement of specific payment measures in in-app billing. As a follow-up to this, an amendment Cabinet order was enforced on March 15, 2022 (Korea Communications Commission (KCC) (2022), NEWS RELEASE, STATE COUNCIL PROVIDES RESOLUTION ON ENFORCEMENT DECREE OF THE TELECOMMUNICATIONS BUSINESS ACT, PROHIBITING FORCING IN-APP PAYMENTS, March 8, 2022.)

In South Korea, despite this kind of law amendment, there appear issues with expensive fees when using an external payment system, and it is speculated that this is why this report in Japan necessitates the *at no cost* part.

[16] Supra note 12, Secretariat of the Headquarters for Digital Market Competition Cabinet Secretariat (2023a).

Most of the report, spanning 192 pages, delves into competitive policy discussions related to the mobile ecosystem. However, here, the scope will be limited to the issue of app stores, outlining the debate with three points of discussion: 1) Are ex-ante regulations necessary in app store regulations? 2) Is government intervention necessary for app store fees? 3) Can the matter of security serve as a justifiable reason for obliging sideloading in app stores?

2.1 Supporting Opinions

Professor Daisuke Korenaga who engaged in the Digital Market Competition Council, and Associate Professor Kazuhiko Fuchikawa, whose expertise are economic law, expressed an affirmative opinion toward government policy with regard to 1) and 2). With regard to 3), Korenaga stated that the involvement of mobile OS companies is essential. Fuchikawa stated that justification due to security matters should be allowed as an exception.[17] Company groups such as the Mobile Content Forum and the Japan Association of New Economy are likely to include many companies that provide content on app stores. They expressed an affirmative opinion toward government policy with regard to 1) and 2). With regard to 3), they stated that there should be a careful decision in relation to whether there are other possible, less restrictive means that will achieve the aims of mobile OS companies.[18]

2.2 Opposing Opinions

Attorney Masayuki Atsumi stated that, with regard to 1), the intervention of private companies through ex-ante regulations should not be allowed without legislative facts, and the Antimonopoly Act should in principle be applied. With regard to 2), he stated that whether digital platforms will adopt the vertical integration model in their business models depends on the company's strategic considerations. With regard to 3), he stated that it was valuable for the report to clarify what is necessary to ensure security.[19]

A professor emeritus of Kansai University's Faculty of Law (full name is concealed in the public comments) stated that, with regard to 1), there is the danger of apps with security risks being downloaded on smartphones. There is an increase in the risk of users falling victims to cyberattacks, in addition to the danger that Apple and Google will be unable to recover app store operation investments. He also noted that this would amount to overregulation that extends beyond individual cases. With regard to 2), he

---

[17] Daisuke Korenaga (2023), Current State of Competition Policy (Part Three), Deep Look in the Smartphone OS Two-Company Monopoly, Nihon Keizai Shimbun Morning Issue, September 21, 2023, Kazuhiko Fuchikawa (2023), A Study of the Competition Assessment of the Mobile Ecosystem Final Report, Fair Trade No. 875, p. 70.

[18] The Mobile Content Forum (2023), Opinions Regarding the Competition Assessment of the Mobile Ecosystem Final Report, 18 August 2023, Japan Association of New Economy (2023), Opinions Toward the Competition Assessment of the Mobile Ecosystem Final Report, August 18, 2023.

[19] Masayuki Atsumi (2023), Study Related to the Competition Assessment of the Mobile Ecosystem Final Report, NBL No. 1248, p. 75.

indicated that having the government decide on the value of app store fees instead of entrusting it to competition amounts to overregulation. With regard to 3), he declared that although the report indicates that requirements for justifiable reasons should be as clear as possible, it should not be limited to the matter of security. For instance, even if it is not justifiable from a security viewpoint, it may be justifiable from the viewpoint of consumer benefit.[20]

With regard to 1), consumer groups such as Consumers Japan and the Japan Association of Consumer Affairs Specialists stated that even if multiple options related to app stores are suggested, consumers will be unable to choose properly and will be exposed to danger. With regard to 2), they expressed concern that it is unclear whether prices will come down, because security measures are necessary. With regard to 3), they indicated that reduced level of security to allow sideloading in app stores will lead to a deterioration of the service.[21]

2.3 Opinions of Those in Charge of the Secretariat of the Headquarters for Digital Market Competition Cabinet Secretariat

Deputy Manager, Tatsuji Narita of the Secretariat of the Headquarters for Digital Market Competition Cabinet Secretariat explained the following in an interview.[22]

With regard to 1), the dominance of OS companies cannot be overlooked. New regulation is necessary for dealing with them. From the viewpoint of developers, there are many problems, not only that of expensive non-transparent app fees, but also content expressions being entrusted to specific companies overseas and the lack of a fair competitive environment in terms of nurturing domestic startups in the digital field. From the viewpoint of consumers, consumers could end up losing options other than the services provided by companies providing OS without them realizing it, suddenly finding themselves being made to buy expensive items. Therefore, ex-ante regulations are necessary.

With regard to 2), concerning the problem of fees being too high, the conclusion reached was that competition will not progress just by opening the payment measures of app stores. In South Korea, enforcing specific payment measures on in-app billing was prohibited in 2021, opening payment measures. Yet, competitive pressures toward fees have not been substantially working properly, as a 26% fee is collected when using external payment systems due to various conditions given by Apple. Therefore, competition between app stores is necessary.

---

[20] Secretariat of the Headquarters for Digital Market Competition Cabinet Secretariat (2023b), Attachment Details of Collected Opinions, Digital Market Competition Council Working Group, No. 52, Distributed Materials 2 (October 12, 2023).

As the same opinion, Toshiaki Takigawa (2024), The Problems of Mobile Ecosystem and App Store regulation, Kokusai Shōji Hōmu Vol. 52 No.1 p. 1.

[21] Consumers Japan (2023), Opinions on the Competitive Assessment of the Mobile Ecosystem Final Report, 9 August 2023, Japan Association of Consumer Affairs Specialists (2023), Opinions on the Competitive Assessment of Consumer Affairs Specialists, August 16, 2023.

[22] Masahiro Sano (2023), Directly Asking People in Charge of the Cabinet the Reasons Why the Government Wants to Make *Sideloading* on Smartphones Mandatory, ITMediaNews, August 22, 2023.

With regard to 3), there is a need to ensure that reliable app stores can compete with each other. Mandatory sideloading will be a part of the ex-ante regulation, and Apple will select a plan allowing the usage of external app stores under co-regulation. Guidelines will be developed for the reviewing of app stores.

3 Considerations

When looking at the individual cases of the EU, the US and Japan, we find the following issues. In the EU, there was the issue of abuse of a dominant position based on Article 102 of the TFEU. In the US, there were issues relating to Restraint of Trade based on Article 1 of the Sherman Act, Monopolization based on Article 2 of the Sherman Act, and violations of the UCL. In Japan, the Antimonopoly Act's Private Monopolization or Trading on Restrictive Terms can be problematic (this was observed when the Fair Trade Commission inspected Apple). Regarding ways to evaluate security issues, the US perceives them as an issue of market definitions and illegality in a two-sided market. They follow an approach wherein the defendant gives proof a procompetitive justifiable reason for their own conducts (the justifiable reason as security measures) in the three-step test. Currently, in Japan, this kind of framework for evaluating market definitions/illegality in a two-sided market has not been established. When considering the justifiable reason of security measures in the Antimonopoly Act in Japan, it should at least be limited to being understood as an exceptional matter, but this may come down to an evaluation of whether it falls under justifiable reasons in terms of Private Monopolization or Unfair Transaction Practices.[23]

When looking at the new legislation, or proposals for new legislation, concerning ex-ante regulations in the EU, the US and Japan, the DMA in the EU, bill for an Open App Markets Act in the US and Japan's suggestion of a new legislation in the Competition Assessment of the Mobile Ecosystem Final Report by the Digital Market Competition Council set up under the Japanese Cabinet contain the provisions of ex-ante regulations related to the three points of 1) mandatory use of payment/billing systems in app stores of digital platforms, 2) the restriction of outlinks, and 3) the restriction of sideloading. In Japan, the Transparency Act specifies Apple and Google as app store-operating companies, and reviews have been carried out. However, as *co-regulation*, this law entrusts companies with independent responses. Entrusting a market-dominated company with independent responses already has its limits as a framework. In this sense, the Digital Market Competition Council's Competitive Assessment of the Mobile Ecosystem Final Report can be appreciated for containing Japan-specific and comprehensive details as a proposal for ex-ante regulations, considering regulation trends overseas.

---

[23] Supra note 17, Fuchikawa (2023), Supra note 8, Tosa (2023).

   Fuchikawa (2023) stated that if justifiable reasons are not understood to a limited extent as exceptions, then a case of carrying out simple balancing may present difficulties in deciding on the superiority or inferiority between different values, resulting in an arbitrary decisions. Moreover, if fair, appropriate competition is not carried out due to security reasons, there will be less range in options for not only third parties, but consumers as well. Furthermore, the digital platform companies should bear the responsibility of giving proof for justifiable reasons related to security.

Finally, in the case of Japan, the Fair Trade Commission inspected Apple and ended its inspection when Apple revised their stipulations of guidelines. However, if one actually confirms the reader apps on a smartphone, it is still sometimes not possible to conduct billing on apps on a smartphone (possibly because of fees by mobile OS companies). Moreover, in the case of billing through payment measures of mobile OS companies, it is still possible to incur a high price upon considering the fee portion. Due to Apple's guidelines restricting the information dissemination to users, there must be many consumers who do not know that a direct contract with a developer without going through Apple's payment measures results in a cheaper fee. Mobile OS companies are asserting the issue of security, but justifiable reasons are only accepted in extremely exceptional situations in terms of the Antimonopoly Act in Japan. Simplistic refutations should not be accepted, and avoidance of responsibility should not be acknowledged. In the domain of desktop PCs, OS companies are already made to carry out security measures in advance. The mobile OS companies' reluctance to carry out measures to security issues while they acquire monopolized benefits in app store operation may have the danger of shifting costs toward consumers in terms of security measures, resulting in acquiring doubled monopolization profits.

In Japan, there is a demand to maintain a fair competitive environment and raise app store competitiveness through the new legislation. If a chance to provide a new kind of app store arrives, this may be the beginning of a revolutionary service.

Note that after writing this paper, Apple announced in a press release on January 25, 2024 that it will offer new options for payment processing and app downloads on iOS within the EU in order to comply with the DMA. Furthermore, on April 26, 2024 the Cabinet approved the draft Law on the Promotion of Competition in Relation to Specified Software Used in Smartphones, which was submitted to the Parliament.

This work was supported by JSPS KAKENHI Grant Number JP22K01291.

References

1. CMA: The Mobile ecosystems Market study final report, 10 June 2022
2. Consumers Japan: Opinions on the competitive assessment of the mobile ecosystem final report, 9 August 2023
3. Korenaga, D.: Current state of competition policy (part three), deep look in the smartphone OS two-company monopoly, Nihon Keizai Shimbun Morning Issue, 21 September 2023
4. Department of Commerce: Competition in the mobile application ecosystem, February 2023
5. Epic Games Inc. v. Apple Inc.: Case No. 4:20-cv-05640-YGR, N.D. Cal., 10 September 2021
6. Epic Games Inc. v. Apple Inc.: Case No. 4:20-cv-05640-YGR, United States Court of Appeals for the ninth circuit, 23 April 2023
7. Epic Games Inc. v. Google LLC.: Case No. 3:20-cv-05671-JD, N.D. Cal., 11 December 2023
8. European Commission: Press release, Antitrust: commission opens investigations into Apple's App Store rules, 16 June 2020
9. European Commission: Press release, Antitrust: Commission sends Statement of Objections to Apple clarifying concerns over App Store rules for music streaming providers, 28 February 2023
10. Fair Trade Commission: Regarding the processes of the suspected antimonopoly act violation case regarding Apple Inc., 2 September 2021

11. Japan Association of Consumer Affairs Specialists: Opinions on the competitive assessment of consumer affairs specialists, 16 August 2023
12. Japan Association of New Economy: Opinions toward the competition assessment of the mobile ecosystem final report, 18 August 2023
13. Fuchikawa, K.: A study of the competition assessment of the mobile ecosystem final report, Fair trade no. 875, p. 70 (2023)
14. Tosa, K.: Evaluation of mandatory in-app billing, the banning of outlinks, etc., in terms of the antimonopoly act, NBL no. 1240, p. 90 (2023)
15. Tagane, K.: Regarding the summary of the competition assessment of the mobile ecosystem final report, Fair trade no. 875, p. 63 (2023)
16. Korea Communications Commission (KCC): News release, state council provides resolution on enforcement decree of the telecommunications business act, prohibiting forcing in-app payments, 8 March 2022
17. Sano, M.: Directly asking people in charge of the cabinet the reasons why the government wants to make *Sideloading* on smartphones mandatory. ITMediaNews, 22 August 2023
18. Atsumi, M.: Study related to the competition assessment of the mobile ecosystem final report, NBL no. 1248, p. 75 (2023)
19. Ministry of Economy, Trade and Industry: Points on the act on improving transparency and fairness on specific digital platforms (Established on 27 May 2020, Distributed on 3 June 2020)
20. Ministry of Economy, Trade and Industry: Evaluation of the transparency and fairness of specific digital platforms (General item-selling online mall and app store fields), 22 December 2022
21. Ministry of Economy, Trade and Industry: 2023 Fiscal year monitoring meeting related to the transparency and fairness of digital platforms, solicitation of opinions [General item-selling online mall/app store fields], 5 December 2023
22. Secretariat of the Headquarters for Digital Market Competition Cabinet Secretariat: Competition assessment of the mobile ecosystem final report, 16 June 2023
23. Secretariat of the Headquarters for Digital Market Competition Cabinet Secretariat: Attachment details of collected opinions, digital market competition council working group, no. 52, Distributed Materials 2, 12 October 2023
24. Hayashi, S.: The antimonopoly act and the banning of outlinks by Apple Inc., TKC Law Library, New Judicial Precedent Commentary watch ◆Economic Law No. 80 Document Number z18817009-00-120802110 (2022)
25. The Mobile Content Forum: Opinions regarding the competition assessment of the mobile ecosystem final report, 18 August 2023
26. Takigawa, T.: The problems of mobile ecosystem and app store regulation. Kokusai Shōji Hōmu **52**(1), 1 (2024)

Machine-Learning Phishing Detection Model Used in the E-Banking Environment

Malvern Manala and Joey Jansen van Vuuren$^{(\boxtimes)}$ (iD)

Department of Computer Science, Tshwane University of Technology,
Pretoria, South Africa
212418625@tut4life.ac.za, JansenVanVuurenJC@tut.ac.za

Abstract. The exponential expansion of Internet usage has given rise to a significant upsurge in cyberattacks, which have caused damage to brand reputation, privacy and financial information, and identities. Phishing, an enduring cyber threat, has emerged as a substantial concern because it can cause considerable financial detriment to economies and erode users' confidence in e-commerce and online banking. The purpose of this research is to identify phishing websites through the development of a Phishing URL Detection Model (PUDM) utilising machine learning techniques. The model utilises machine-learning algorithms and a standard UCI machine-learning library dataset. The performance of the proposed methods surpassed that of several machine learning algorithms when compared to related work for this study. A feature importance analysis was performed to ascertain the features that would be employed to distinguish phishing URLs from authentic ones. The study determined that the Google Index feature had the greatest impact on determining the validity of website URLs. In contrast, the XGBoost classifier demonstrated the highest performance, attaining an F1 score of 92.72%. The results of this study have the potential to significantly enhance the security measures in place for organisations, clients, and website proprietors.

Keywords: Phishing · Cybercrime · Machine Learning · Website URL · e-Banking

1 Introduction

Since its inception, virtually everyone now utilises the Internet for virtually any purpose. New York central banks pioneered the implementation of e-banking services in 1981. Customers can use the Internet to perform financial transactions and utilise banking services through e-banking, also known as online banking or virtual banking. E-banking enables clients to manage and conduct account transactions with the bank electronically without the need for physical presence at the branch. Financial transactions can be executed through Internet banking on either a desktop computer or a mobile device. The majority of financial institutions provide customers with e-banking alternatives. Banks employ e-banking, a web-based application, to conduct online banking. It comprises software applications that are accessible via a web browser, such as the Capitec online

© IFIP International Federation for Information Processing 2024
Published by Springer Nature Switzerland AG 2024
R. M. Davison and D. Kreps (Eds.): HCC 2024, IFIP AICT 719, pp. 69–85, 2024.
https://doi.org/10.1007/978-3-031-67535-5_7

banking application of the Capitech bank in South Africa. Online banking is emerging swiftly as a disruptive force within the perpetually expanding financial services domain. The progression of information technology (IT) has enabled financial institutions to transition from conventional banking practices to modern, technology-driven alternatives. The proliferation of e-banking is anticipated to increase substantially due to the Internet's popularity. Users can conveniently and expediently carry out financial transactions remotely from their homes or places of business through e-banking [1]. At present, patrons of financial institutions are granted the capability to conduct transactions electronically via e-banking, in contrast to the traditional approach of physically visiting the bank to solicit services and satisfy their needs [2].

One of the main limitations of e-banking is its inadequate security measures. Cybercriminals are using malicious software and presenting erroneous transactions that have not been executed to target e-banking users. Phishing has become the prevailing method of targeting e-banking customers due to its straightforward implementation process. Upon conducting a transaction on a fraudulent website, an individual's login credentials and sensitive data are compromised without delay. Social engineering attacks are purposeful and deceitful activities that deceive users into divulging sensitive information by convincing them they are making a security-related choice [3]. Furthermore, as stated by Wang, Zhu [3], "social engineering" is a form of cyberattack in which the assailant exploits human susceptibilities to compromise the security objectives of cyberspace components—including audibility, controllability, confidentiality, integrity, and availability—through the use of deception, manipulation, coercion, influence, persuasion, data, resources, users, and operations. Phishing represents a form of social engineering scheme wherein the perpetrator masquerades as a reputable website in order to acquire sensitive data [4]. E-banking phishing is a type of cyber assault wherein the attacker integrates a spoof website into the legitimate site in order to acquire sensitive banking information from the user, including their bank PIN, username, and password [5]. As the number of Internet users continues to rise, phishing attempts pose an increasingly serious security risk to society.

Phishing is a global phenomenon; Table 1 below shows the World Cybercrime Index (WCI) score for the top 15 countries that have participated in a survey conducted by [6].

As noted in Table 1, from March to October 2021, Bruce and Lusthaus [6] invited experts in cybercrime intelligence and conducted an anonymous online survey to identify the geographical locations of cybercrime offenders. Bruce and Lusthaus [6] identified five major categories of cybercrime, named countries as significant sources, and ranked them based on their impact, professionalism, and technical skills. The WCI, a global metric, revealed that a few countries house the greatest cybercriminal threats.

Phishing attacks constitute the source of at least 90%t of all successful hacking techniques. According to the findings of the 2017 IBM Threat Intelligence report, the volume of phishing emails has doubled within the past few years [7]. Approximately 50% of the 15 billion spam and phishing emails that are sent daily attempt to impersonate or target financial institutions [7]. Although many organisations have implemented spam filters that effectively identify and obstruct 99% of unsolicited emails, the most dangerous variants of spam generally possess the most advanced technological features [8]. According to the BusinessTech survey that quantifies customer satisfaction with digital

Table 1. World Cybercrime Index overall–top 15 countries [6].

Rank	Country	I	P	TS	WCI Score	Tech	Attacks	Data	Scams	Cash
1	Russia	8.96	8.81	8.73	58.39	82.17	81.34	65.18	21.70	41.56
2	Ukraine	8.37	8.29	8.24	36.44	52.97	50.76	36.01	11.20	31.27
3	China	8.22	7.70	7.81	27.86	40.22	24.24	34.89	15.83	24.13
4	United States	7.99	7.21	7.21	25.01	27.64	17.68	30.36	22.72	26.63
5	Nigeria	8.25	6.49	5.80	21.28	7.93	8.41	23.04	52.17	14.86
6	Romania	7.12	7.04	7.15	14.83	17.83	9.17	22.50	13.15	11.49
7	North Korea	7.91	7.23	7.38	10.61	8.66	25.33	13.01	2.17	3.88
8	United Kingdom	7.86	7.21	6.75	9.01	5.04	4.75	5.80	7.86	21.63
9	Brazil	6.90	6.35	6.32	8.93	13.70	8.77	10.29	7.28	4.64
10	India	7.90	6.60	6.65	6.13	4.46	3.62	6.81	12.75	3.01
11	Iran	6.88	6.45	6.64	4.78	8.62	10.00	3.59	0.94	0.72
12	Belarus	6.84	7.20	7.32	3.87	11.92	5.58	1.85	--	--
13	Ghana	8.57	6.83	6.09	3.58	1.23	0.76	2.97	10.36	2.57
14	South Africa	6.95	5.35	5.50	2.58	1.20	0.65	0.58	7.17	3.30
15	Moldova	7.38	7.19	7.56	2.57	6.70	0.98	2.43	0.83	1.88

I = Impact; P = Professionalism; TS = Technical skill, Technical = *Technical products/services*, Attacks = *Attacks and extortion*, Data = *Data/identity theft*, Cash = *Cashing out and money laundering*. I, P, and TS are scored out of 10. 'WCI Score', and all columns following, are scored out of 100. Each country's top score across all cybercrime types is shaded in grey.

banking services in South Africa, customers are less satisfied with digital banking services in South Africa [9]. The report also revealed that the rate of fraud has decreased over the past few years, currently standing at 46%, down from 50% in 2016, 55% in 2015, and 62% in 2014. However, the number of consumers experiencing losses has increased, with 12% affected in 2015, 14% in 2016, and 19% in 2017 [9]. Since 2017, fraud increased exponentially, and the 2021 results showed that 37% of South African customers were targeted with digital scams, and 29% lost money due to that [10]. In 2016, Kaspersky Labs found that 1.09 million instances of banking Trojan infections were detected during 2016, representing a 30.6% increase compared to the previous year [11]. According to the findings of their investigation, around 47.48% of phishing attacks involved the redirection of users to illegitimate banking websites or pages with the intention of acquiring login credentials. Since the majority of phishing sites are concealed within legitimate domains, security organisations face a challenge in analysing the compromised phishing data and developing techniques to detect the breach [12]. Kaspersky [13] also indicated that the percentage of users impacted by phishing in South Africa already increased to 9.7% of the South African users. The most frequently targeted sectors with financial phishing attacks were e-commerce platforms and online financial institutions. Faux payment system websites accounted for 15.4% of financial phishing attempts in South Africa, while 68.4% were conducted through bogus online stores and 16.2% were through illegitimate online bank portals. South African Banking Risk Information Centre (Sabric) released its Annual Crime Stats 2022, showing a 24% rise in reported cases of digital banking fraud compared to the previous year. The increase was mainly due to the rising number of fraud cases associated with banking applications and e-banking [14]. Ensuring the protection of electronic banking customers from these types of attacks is of the utmost importance, given the significant financial losses they can cause.

This research focused on the development of a machine learning method for identifying and validating phishing websites utilised in e-banking attacks. The implementation of supervised machine learning methods to identify fraudulent URLs decreases the vulner-ability of e-banking users to phishing attacks. A classification machine learning model is utilised to forecast the categorisation of phishing URLs. The proposed Phishing URL Detection Model (PUDM) would evaluate the attributes of the uniform resource location (URL) to determine whether it is legitimate or phishing. A validation process will be performed on the page before granting users access or facilitating real-time online trans-actions. Phishing URLs can be effectively detected using machine learning techniques, given the transient nature of phishing websites [15]. However, conventional techniques like blocklist and allowlist fail to detect phishing websites on time, do not possess pre-dictive capabilities, and necessitate the initial URL reporting on phishing databases like PhishTank.

Subsequently, machine learning models can also be used to refine the updated dataset. The machine learning model developed as part of this study can be used to evaluate URLs in real-time. The dataset that was provided showed that the proposed methods performed superiorly to alternative techniques. Feature importance analysis is conducted using the provided dataset and the proposed solution to evaluate the basis for the machine learning decision-making and examine which attributes are normally employed to determine URLs as either phishing or authentic. The Google indexing function exerted the most substantial influence in ascertaining the legitimacy of a website. XGBoost demonstrated the highest level of performance, attaining an F1 score of 92.72% and an accuracy rate of 97%t. This proposed solution can substantially enhance online banking safety for clients, organisations, and proprietors of websites.

XGBoost models have been employed in a variety of fields, ranging from finance to health and beyond, to show remarkable performance. For example, the health sector has an aspect where xgboost is at work for early diagnosis, disease prediction, and drug discovery. In marketing and eCommerce, churn prediction and recommendation systems have been implemented, while XGBoost models have been used for customer segmentation. Hence, XGBoost has shown a boost in the field of fraud detection and cyber security, having the ability to discriminate normal behaviour and detect fraudulent activity.

2 Related Work

A literature survey was done on phishing detection to determine how others solved some of these problems, with a focus on how these methods can be improved. The following detection methods were studied:

2.1 List-Based Approach

Using an auto-update whitelist, Jain and Gupta [16] proposed an innovative method for protecting against phishing attacks on the client side. The true positivity rate of their solution is 86.02%, while the false positivity rate is 1.48%. Similar to most list-based methods, a phishing URL is labelled with a status to indicate that it is a phishing URL, and it is then stored as such in a database. Consequently, this method is simpler to implement, as it is possible to identify phishing by checking if the URL is in the database.

This solution's limitation is that minor URL modifications will result in non-detection by this list-based detection solution. [16].

2.2 Visual Similarity Approach

Cernica and Popescu [17] employed JavaScript hash functions to identify similarities among multiple pages and compared the computer vision to other websites based on the login structure of the request. The solution is provided via a browser plug-in that establishes a connection with a server containing a machine-learning algorithm.

2.3 Heuristic Rule Base Approach

This blacklist extension is capable of detecting new attacks by identifying phishing URLs according to the characteristics of the phishing site. This restriction might be inadequate in identifying every new attack, as attackers can bypass the identification if they know the methods and functions used. Furthermore, the effectiveness of this approach can be compromised due to a potential absence of the common characteristics on the website. An alternative methodology entails scrutinising the payloads of different protocols utilising applications on the client or server where the arbitrary protocols were used [18]. Machine learning algorithms can be used to identify phishing attacks and, therefore, strengthen systems' security. It is possible to identify zero-hour phishing attacks (new attacks that were not seen before) using a series of general heuristic checks [19]. This method offers a distinct advantage as it can overcome the restrictions of the blacklist methodology. Teraguchi and Mitchell [18] indicated that it is important to note that blacklists require exact matches, and consequently, precise attacks must be observed prior to their inclusion in the blacklist. The problem is that generalised heuristics may incorrectly classify valid content, including emails or websites that do not fall under the spam category. According to Teraguchi and Mitchell [18], heuristic protections are deliberately integrated into widely used email clients and web browsers in order to detect phishing attacks. Anti-phishing heuristics may also be applied as a ClamAV10-like approach. In addition, there is also a plug-in developed by Stanford University that can detect HTTP-based phishing attempts [18].

2.4 Machine Learning Approach

The methodology used in this study entails the extraction of distinctive characteristics, followed by the implementation of machine learning algorithms to perform classification. In order to classify phishing URLs, common datasets of the URLs and associated websites are gathered, which include the URL details, website architecture, and JavaScript functionalities.

Phishing is the result of a variety of motives, such as monetary benefits, data theft and identity theft. Attackers use human psychology and physical vulnerabilities in digital systems to steal sensitive data, such as financial information and login credentials [28]. Phishing attacks are usually associated with social engineering, malware distribution, human behaviour and system weakness exploitation. Phishing is curtailed via a multi-pronged strategy that incorporates user training, cyber awareness and technological tools

[28]. Traditional phishing detection approaches have a number of limitations that prevent them from effectively fighting against the increasing number of phishing attacks. These weaknesses are in the form of static signatures dependency, limitation in the detection of novel threats, high false positive rates, lack of ability to detect polymorphic attacks, limited behavioural analysis, manual review process reliance, zero-day exploit detection difficulty, and real-time detection lack. Such disadvantages also point out the necessity to develop the most sophisticated and dynamic techniques for phishing detection to mitigate the threat of phishing attacks in modern times.

Following this, phishing datasets are obtained through the utilisation of the afore-mentioned characteristics. These features are used in the training of machine learning algorithms to detect phishing sites.

Rashid, Mahmood [20] proposed a machine-learning approach that uses Alexa and the Common Crawl archive dataset with 5000 URLs. The accuracy of this proposed approach to classify phishing and legitimate websites is indicated as 95.66%. Rashid, Mahmood [20] employed the support vector machine approach to train the model. The study can be seen as shallow as it has used only a singular classifier, namely the Support Vector Machine (SVM), together with a limited set of only five features for identifying phishing websites. A small set of datasets was gathered utilising GNU and Python scripts. In addition, the evaluation of the model was solely based on only one performance indicator, namely accuracy.

Geyik, Erensoy [21] introduced a Decision Tree machine-learning methodology. The machine learning algorithms used are logistic regression, Naive Bayes, and Random Forest. Geyik, Erensoy [21] used the PhishTank Alexa Common-Crawl datasets. The performance attained using the dataset is much lower when compared to previous studies employing similar classifiers and datasets. The Random Forest classifier only had an accuracy of 83.0%, and there was no report of an F1 score.

Korkmaz, Sahingoz [22] also used a machine-learning approach. The model used logistic regression, K-nearest neighbour, Decision Tree, support vector machine, Naive Bayes, XGBoost, Random Forest, and artificial neural networks. The dataset PhishTank Alexa Common-crawl was used. The Random Forest algorithm using dataset 1 yielded the highest accuracy, with a rate of 94.59%. The performance in this study is much lower when compared to other studies employing similar classifiers and datasets.

The study conducted by Patil, Thakkar [23] utilised a limited number of commonly used techniques for model evaluation as well as a smaller dataset compared to the other research. While evaluating the results, the study demonstrates minimal instances of false-positive and false-negative outcomes. Patil, Thakkar [23] used a dataset from Alexa consisting of 9076 websites with a machine learning methodology that included Decision Trees, logistic regression, and Random Forest algorithms. The Random Forest classifier in the study had the best accuracy (96.58%).

This paper demonstrates that using diverse features and knowledge from prior research can lead to greater accuracy. In contrast to previous studies, this paper was carried out to evaluate advanced metrics such as the F1 score measure to obtain a balanced evaluation of the model's performance. The feature importance technique was used to analyse URLs with 21 features. The comparative analysis was used to assess the accuracy rates, F1 scores, and model training times of the various algorithms using the machine-learning approach.

3 Phishing URL Detection Model (PUDM)

3.1 The URL

The objective of this paper is to develop a phishing URL detection model through the analysis of webpage URLs. A URL is a sophisticated sequence of characters that represents both the structure and meaning of a resource accessible on the Internet [24]. Figure 1 depicts the basic structure of a URL.

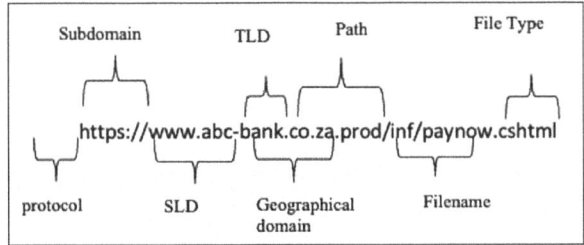

Fig. 1. The URL Structure [29]

A URL always starts with the name of the access protocol for a webpage. After that, the top-level domain (TLD), which indicates the domains in the DNS root zone of the Internet, is identified. It is followed by the subdomain and the second-level domain (SLD) name, which generally corresponds to the organisation name or the hosting server.

3.2 Dataset

In this paper, an open-source dataset was obtained from the Kaggle dataset library. This dataset contains legitimate sites from the Alexa database and phishing sites from PhishTank. The dataset was developed to serve as a benchmark for phishing detection systems using machine learning. The collection is evenly distributed, with a split of 50% phishing URLs and 50% legal URLs [25]. Figure 2 is a bar graph representing the features of the dataset. It displays the distribution or counts of the "Status" label class (0 and 1) for each feature in the dataset. The legitimate ('0') website is represented by blue, and the phishing website is characterised by brown.

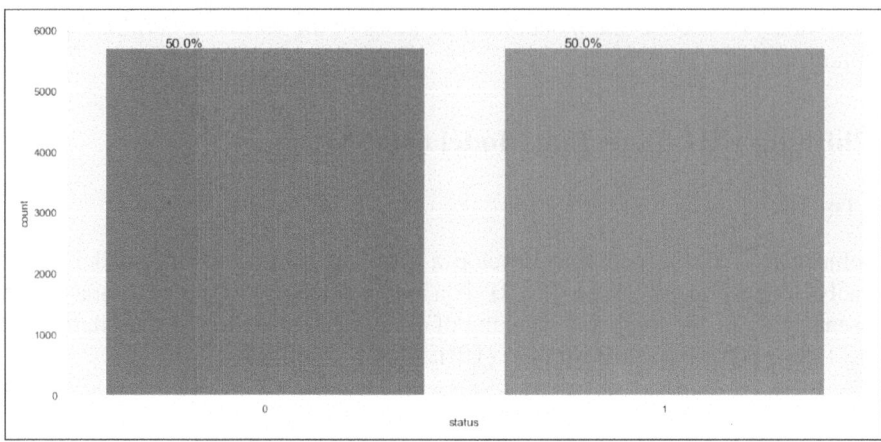

Fig. 2. Phishing Dataset distribution

3.3 Feature Selection

Random Forests is a Bagging Algorithm that combines a specified number of Decision Trees. The Random Forests utilise tree-based techniques that rank nodes based on their ability to enhance purity by reducing impurity (Gini impurity) across all trees. The nodes with the most significant reduction in impurity are located at the beginning of the trees, whereas nodes with the smallest reduction in impurity are found at the end of the trees. Pruning trees below a specific node allows us to generate a subset of the most important features.

Following the Random Forest Feature importance algorithm, the arrangement of the features was implemented, and the top 20 features that would be employed were determined. Additionally, Python correlation coefficient libraries were employed to evaluate feature importance and identify the top 20 highly predictive features. The objective of the study was to attain a greater level of accuracy in terms of the F1 score. The attributes used in this study are listed in Table 2.

Table 2. The summary features a description of our dataset

Feature	Description	Datatype
Features extracted from Address Bar		
Number of at (@)	The URL contains the character "@" everywhere and the number of @ in a URL	Number
Number of Colons	Several times, the Colon symbols are found within a URL	Number
Number of slashes	Several times, the "/" slashes symbols are found within a URL	Number

(continued)

Table 2. (*continued*)

Feature	Description	Datatype
Number of and (&)	The number of times the "%" symbols are found within a URL	Number
Number of semicolons	The number of times the Colon symbols are found within a URL	Number
Number of spaces	Presence of space within a URL	Number
The number of dots(.)	Checking whether there is more than one "." in the URL route can quickly reveal if it's phishing or not	Number
Number of a percent (%)	The number of times the "%" symbols are found within a URL	Number
Number of underscores	The number of times the "_" symbols are found within a URL	Number
Length of the URL	The length of a URL	Number
IP Address	URLs that use IP addresses rather than domain names. Some IP addresses undergo hexadecimal coding transformation	Boolean
Number of subdomains	The number of subdomains extracted with the URL	Number
Prefix suffix	Use of the dash (-) sign-in URLs to make a website appear authentic	Boolean
HTML and JavaScript feature-based		
Popup window	Use of a pop-up window	Boolean
Domain Base Features		
DNS record	No records were found for the hostname in the database or identity that was not recognised in the WHOIS database	Boolean
Web_traffic	The feature was taken from the Alexa database evaluating website popularity. Check to see if the domain receives no traffic or is not listed in the Alexa database	Number
Google Index	Determines whether or not Google has indexed the website	Number
Abnormal Based Features		
SFH	Determines whether the domain name in SFHs is different from the domain name of the webpage or if the Server Form Handler (SFH) contains empty text or "about blank."	Boolean

3.4 PUDM Implementation

This study experimented with training seven different algorithms in a machine learning-based model. These included Gradient Boosting (XGBoosting), Stacking Classifier, Random Forest (RD), Decision Trees (DT), Tuned Decision Tree (TDT), Adaptive Boosting (AdaBoost), and Bagging Classifier. The models generated by these algorithms were trained using the Scikit-learn library in the Python programming language. The following machine-learning algorithms were used for the study:

Decision Tree is a data structure that divides data into subsets based on specific conditions related to the features [26]. At each node, decisions are made, resulting in a structure resembling a tree. The tree form facilitates the representation of complex connections within the data. Tuned Decision Tree uses features to make decisions by dividing the data into subsets. Tuning entails the process of adjusting hyperparameters, such as tree depth and minimum samples per leaf to enhance the algorithm's performance. An optimally tuned Decision Tree achieves an optimal balance between complexity and predictive accuracy.

Random Forest algorithm develops numerous Decision Trees using random selections of the input [26]. The ultimate prediction often involves aggregating the individual tree predictions through an average or voting process. By integrating a variety of models, it reduces overfitting and enhances generalisation. XGBoost is a machine-learning technique that employs Decision Trees to form an ensemble model. The process involves iteratively building Decision Trees to optimise a loss function, where each tree is designed to prevent the faults identified by the previous ones [27]. The model includes regularisation terms to manage the complexity of the model and avoid overfitting. AdaBoost Classifier is a machine learning algorithm that places emphasis on training weak models in a sequential manner [26]. By giving incorrectly classified instances larger weights, each iteration gives them more significance. The ultimate model is a composite of weak models, emphasising difficult instances that are complex to classify.

Bagging Classifier Bagging is a technique that involves training numerous instances of the same basic model on subsets of the data that are created using bootstrapping [26]. The ultimate prediction typically involves aggregating the predictions of individual models through an average or voting method. It enhances stability and minimises variance in the model. Stacking Classifier is a technique that involves combining many base models to create a meta-model. The base models generate predictions based on the input data, which are then used as features for the meta-model. The meta-model is trained to generate the ultimate prediction by employing the outputs of the basis models.

In this study, we implemented and trained the DT, TDT, RF, Bagging, AdaBoost, XGBoost, and Stacking Classifier using Python Scikit-Learn machine learning libraries. Moreover, these classifiers were trained using a phishing URL dataset sourced from the Kaggle library (Table 1). DT and RF are the basis for this study. The different model implementations are aimed at comparing performance results and obtaining the best-performing model for the PUDM model. Algorithms like XGboost, Bagging, AdaBoot, and Stacking were aimed at improving and boosting the model's performance for this paper. The results of each model's performance are detailed in the experimental results section of this paper.

As for the generalisation of the results, although our research is confined to phishing URL detection in e-banking, the proposed methodology can be easily customised and used in other areas that have analogous properties. Principles of feature extraction, model training, and performance evaluation are universally applicable to other contexts beyond e-banking. Nevertheless, it should be noted that the models' performance of machine learning models depends on factors like the dataset's characteristics, feature engineering techniques and hyperparameters of the model. So, additional validation and adjustments might be required to use the suggested approach in other domains. However, the robustness and the flexibility of XGBoost models allow for their usage in many applications, and the results will add to the body of knowledge of machine learning-based cybersecurity approaches.

3.5 Hyperparameter-Tuning and Cross-Validation

The purpose of doing cross-validation was to ensure that the hyperparameters of the model were appropriately tuned and produced consistent results when applied to simulated data. While other cross-validation variations exist, we have chosen to utilise K-fold cross-validation for the purposes of this investigation. In K-fold cross-validation, the initial training set (obtained by partitioning the original dataset into a training set and a test set) is divided into N number of folds, as illustrated in K Fold Cross Validation [28] (with N = k folds). One-fold is subsequently utilised as a test set from that fold (or validation set in our case). Following this, the remaining folds are utilised for testing. The initial iteration concludes with the completion of the model testing phase using the current configuration and the specific set of hyperparameters. The second iteration continues with training, using a different test fold for evaluation and the remaining folds with the same set of hyperparameters. This process is repeated N times in order to utilise each fold for an experiment precisely once.

In this paper, each fold's average evaluation metrics and hyperparameters are documented to showcase consistency. A set of hyperparameters that performs well across all folds is demonstrated. To evaluate the metrics' performance, knowledge of their mean value and deviation is required. Following this, the model is trained and evaluated using the same approach with a new set of hyperparameters. The candidate with the lowest deviation and the highest mean across all criteria is selected. The process is automated using GridSearchCV through Python libraries. The researcher used 10-fold cross-validation and GridSearchCV for hyperparameter tuning to improve the model accuracy. The Scikit-Learn library was used to implement the Grid-Search. The experiments were run on a computer with 8 GB RAM and 2 2-core Intel i5 processor. The PUDM model was constructed for every permutation of a specified hyperparameter and then internally assessed and ranked within the provided cross-validation folds (Fig. 3).

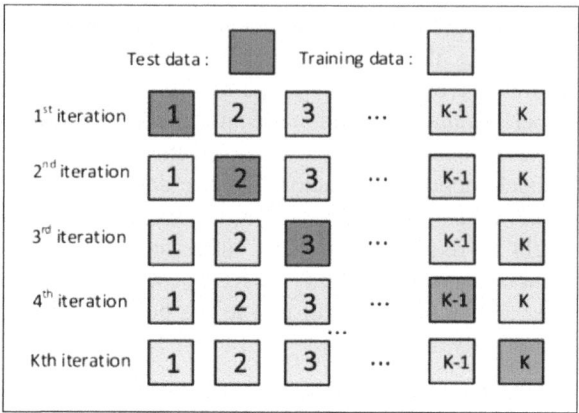

Fig. 3. K Fold Cross Validation [28]

4 Experimental Results

This section outlines the evaluation metrics employed in the study to evaluate the performance of the PUDM model algorithm and presents the results that were implemented.
The experimental results of the PUDM Model algorithms, along with their pros and
cons discussed in the previous section, are employed with the phishing URL dataset.
The experiment shows results in two categories: training accuracy and test accuracy. This
section compares the Receiver Operating Characteristics (ROC) accuracy results of the
PUDM Model algorithms to determine the best model for detecting phishing URLs in
this study.

Table 3. Performance Metrics

True Positives (TP): the data points that a classifier predicts as positive and whose actual value is positive as well True Negatives (TN): Data points are predicted as negative by a classifier whose actual value is negative	False Positives (FP): The data points that a classifier incorrectly predicts as positive while the actual value is negative False Negative (FN): The data points that a classifier predicts as negative when the real data is observed as positive
$Specificity = \frac{TN}{TN+FP}$	$Sensitivity = \frac{TP}{TP+FN}$
$Accuracy = \frac{TP+TN}{P+N}$	$Recall(FAR) = \frac{FP}{FP+TN}$
$Precision = \frac{TP}{TP+FN}$	$F1Score = 2\frac{FP+FN}{P+N}$

4.1 Evaluation Metrics

Model performance in this study was done using the machine learning confusion matrix. A confusion matrix is a tool used to evaluate the performance of a machine-learning model. The confusion matrix for determining the classification model's performance in this study is implemented as an N x N matrix, with N representing the total number of target classes. The matrix puts together the real target values with the ones predicted by the machine learning model. This provides a comprehensive perspective on the performance of our PUDM model and the types of errors it generates. The performance metrics are shown in (Table 3). According to our hypothesis, the data will be classified as positive for phishing websites and negative for authentic websites.

4.2 Comparison of Training vs Test Results

Our phishing detection model was evaluated using a dataset of 11,430 legitimate and malicious URLs. In this sample, there were 5,715 phishing URLs and 5,715 authentic URLs. Normally, phishing traffic contains a significantly smaller number of begin URLs than legitimate traffic. Our findings in (Table 4) suggest that our detection PUDM model maintains its outstanding performance even when working with this balanced dataset. The high level of accuracy indicates that our proposed model classifies URLs as either legitimate or fraudulent with minimal ambiguity.

Moreover, throughout our research, the sensitivity of the model was marginally greater than its precision. This finding suggests that our model exhibits a higher tendency to produce false positives as opposed to false negatives. This represents a reasonable compromise, as frequent false positives resulting from the erroneous identification of legitimate websites as phishing sites would merely annoy users. A higher false negative rate would not be desirable as that will increase the users' vulnerability.

Table 4. Summary of the training and test accuracy results.

Model	Train_Accuracy	Test_Accuracy	Train_Recall	Test_Recall	Train_Precision	Test_Precision	Train_F1-Score	Test_F1-Score
XGBoost Classifier	0.953131	0.927676	0.94200	0.921866	0.963436	0.932743	0.952598	0.927273
Stacking Classifier	0.949006	0.922426	0.94250	0.913703	0.954914	0.929970	0.948666	0.921765
Random Forest	0.944632	0.919802	0.93450	0.915452	0.953815	0.923529	0.944059	0.919473
Tuned Decision Tree	0.936133	0.916302	0.91525	0.899708	0.955127	0.930639	0.934763	0.914913
AdaBoost Classifier	0.906762	0.904345	0.90250	0.904956	0.910237	0.903902	0.906352	0.904429
Decision Tree	0.799650	0.789735	0.97650	0.974344	0.721330	0.711670	0.829740	0.822545
Bagging Classifier	0.499938	0.500146	1.00000	1.000000	0.499938	0.500146	0.666611	0.666796

Figure 4 depicts the confusion matrix for the XGBoost classification algorithm. The XGBoost classifier obtained the highest F1 score against other classification algorithms in terms of URL phishing detection. The XGBoost model achieved the highest F1 Score by enhancing the performance of the Random Forest algorithm. Enhancements were made to the initial results achieved from RF and DT by incorporating advanced techniques such as XGBoost, AdaBoost, and Bagging. The enhancement made with the bagging classifier did not yield the anticipated outcome, resulting in a 67% performance on the F1 Score metric.

The study also aimed to employ a classification algorithm that would improve the way phishing algorithms are detected in terms of prediction. The F1 score findings prove that the classification algorithm can be employed to improve the verification or detection of phishing URLs using machine learning, as the accuracy is higher than other techniques used in the referenced studies.

4.3 Comparison of Model Confusion Metrics Results

The PUDM model confusion matrix shows the counts of true positives (TP), true negatives (TN), false positives (FP), and false negatives (FN). This matrix assisted in evaluating model performance, detecting misclassifications, and enhancing predictive accuracy for identifying phishing URLs. Figure 4 shows the experimental result of the confusion matrix performance on Stacking, Adaboost, XGBoost, RF, Bagging and DT for this paper. The confusion matrix used to compute the obtained outcomes was done using the Python machine learning Scikit-Learn confusion matric library. As depicted, XGboost obtained the highest TP more than any other algorithm during training. These results indicated the precision of the PUDM model.

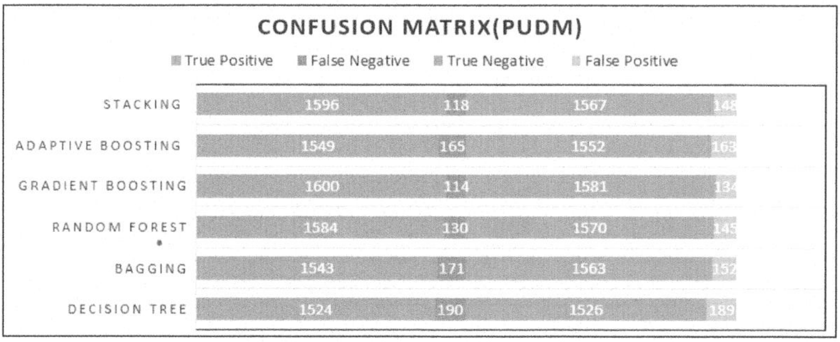

CONFUSION MATRIX(PUDM)

■ True Positive   ■ False Negative   ■ True Negative   ■ False Positive

	TP	FN	TN	FP
STACKING	1596	118	1567	148
ADAPTIVE BOOSTING	1549	165	1552	163
GRADIENT BOOSTING	1600	114	1581	134
RANDOM FOREST	1584	130	1570	145
BAGGING	1543	171	1563	157
DECISION TREE	1524	190	1526	189

Fig. 4. PUDM Model Confusion Metrics

4.4 Comparison of Receiver Operating Characteristics (ROC) Accuracy

The study evaluated the effectiveness of various classification models for phishing detection in e-banking URLs. The main objective was to evaluate and compare the models' ROC accuracy. Figure 5 shows the results of the ROC accuracy for the XGboost, Bagging, RF, Adaboost, DT, and Stacking Classifiers implemented in this study. The ROC was implemented using the Python Scikit-Learn machine learning libraries using the phishing URL dataset.

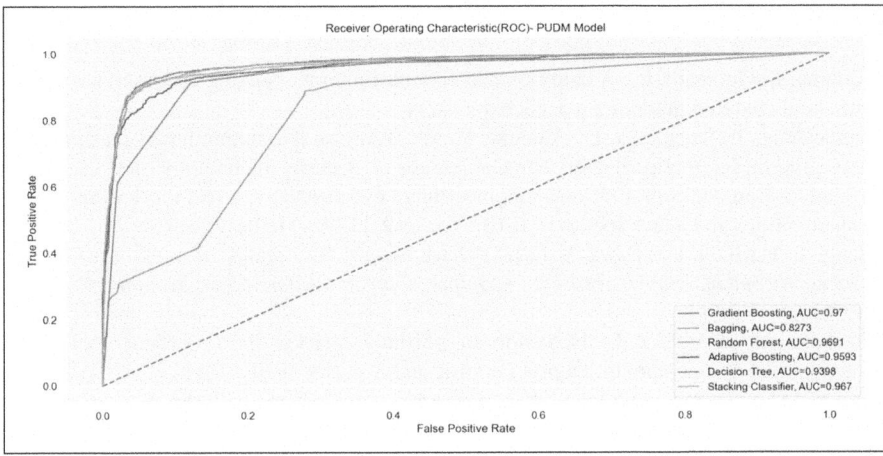

Fig. 5. ROC Accuracy comparison

The bagging model showed the lowest performance in terms of TP, with an accuracy rate of 87%. However, most models demonstrated promising performance in both TP and FP rates, indicating their ability to classify phishing URLs while minimising false alarms. The XGBoost model, which had an accuracy rate of 97%, was found to be the most effective. This is due to its ability to handle categorical data without explicit data encoding. The dataset used in the study consisted of more categorical features than numerical features, highlighting the superiority of the gradient boosting model in accurately detecting phishing URLs in e-banking environments.

5   Conclusion

The objective of this research was to develop a phishing URL detection model, utilising several machine learning algorithms. The proposed model underwent testing using recent datasets from the literature, and the results were compared with both the test and training sets. The comparison of the algorithmic outcomes demonstrates that the proposed approach significantly improves the effectiveness of phishing detection and achieves high accuracy rates. The study has shown that the XGBoost model performed the best and presents a more reliable and robust solution for protecting e-banking users from phishing attacks.

This study can be enhanced by the consolidation of different datasets into a single comprehensive dataset for the purpose of detecting phishing attempts. This will be achieved by employing a big data platform to develop a model specifically designed for detecting phishing URLs. By using this dataset, the objective will be to improve the PUDM through the implementation of deep learning algorithms and explore NLP (natural language processing) techniques for generative AI.

References

1. Kavitha, J., Gopinath, R.: A study on perception of internet banking users service quality-a structural equation modeling perspective (2021)
2. Nathezhtha, T., Sangeetha, D., Vaidehi, V.: WC-PAD: web crawling based phishing attack detection. In: International Carnahan Conference on Security Technology (ICCST) (2019)
3. Wang, Z., Zhu, H., Sun, L.: Social engineering in cybersecurity: effect mechanisms, human vulnerabilities and attack methods. IEEE Access **9**, 11895–11910 (2021)
4. Peng, T., Harris, I. and Sawa, Y. (2018) *Detecting phishing attacks using natural language processing and machine learning. In 12th International Conference on Semantic Computing,* IEEE
5. Baykara, M., Gürel, Z.Z.: Detection of phishing attacks. In: Proceedings of the 6th International Symposium on Digital Forensic and Security. IEEE (2018)
6. Bruce, M., et al.: Mapping the global geography of cybercrime with the World Cybercrime Index. PLoS ONE **19**(4), e0297312 (2024)
7. Loeb, L.: Cybersecurity incidents doubled in 2017, Study Finds (2018). https://securityintelligence.com/news/cybersecurity-incidents-doubled-in-2017-study-finds/
8. Moramarco, S.: Phishing attacks in the banking industry (2019). https://resources.infosecinstitute.com/topic/phishing-banking-industry/
9. Buisnesstech: Big jump in reported digital banking fraud in SA (2017). https://businesstech.co.za/news/banking/176663/big-jump-in-reported-digital-banking-fraud-in-sa/
10. Businesstech: Shock findings on digital fraud in South Africa, and the one scam you should avoid (2021). https://businesstech.co.za/news/it-services/484593/shock-findings-on-digital-fraud-in-south-africa-and-the-one-scam-you-should-avoid/
11. Barth, B.: Kaspersky: banking malware attacks up 30.6% in 2016; finance sector phishing also more prevalent (2017). https://www.scmagazine.com/news/kaspersky-banking-malware-attacks-up-30-6-in-2016-finance-sector-phishing-also-more-prevalent
12. Zhu, E., et al.: An effective neural network phishing detection model based on optimal feature selection. In: IEEE International Conference on Parallel & Distributed Processing with Applications, Ubiquitous Computing & Communications, Big Data & Cloud Computing, Social Computing & Networking, Sustainable Computing & Communications. IEEE
13. Kaspersky: 8.7% of users encountered phishing attacks in Africa in 2022, global number of attacks exceeds 500 million (2023). https://www.kaspersky.co.za/about/press-releases/2023_87-of-users-encountered-phishing-attacks-in-africa-in-2022-global-number-of-attacks-exceeds-500-million
14. Moyo, A.: SA sees alarming rise in digital banking fraud (2023). https://www.itweb.co.za/article/sa-sees-alarming-rise-in-digital-banking-fraud/VgZey7JlzZPqdjX9
15. Gutierrez, C.N., et al.: Learning from the ones that got away: detecting new forms of phishing attacks. IEEE Trans. Dependable Secure Comput. **15**(6), 988–1001 (2018)
16. Jain, A.K., Gupta, B.B.: A novel approach to protect against phishing attacks at client side using auto-updated white-list. EURASIP J. Inf. Secur. **1**, 1–11 (2016)
17. Cernica, I., Popescu, N.: Computer vision based framework for detecting phishing webpages. In: Proceedings of the 19th RoEduNet Conference: Networking in Education and Research (RoEduNet). IEEE (2020)
18. Teraguchi, N., Mitchell, J.C.: Client-side defense against web-based identity theft. Computer Science Department, Stanford University (2020). http://crypto.stanford.edu/SpoofGuard/webspoof.pdf
19. da Silva, C.M.R., Feitosa, E.L., Garcia, V.C.: Heuristic-based strategy for phishing prediction: a survey of URL-based approach. Comput. Secur. **88**, 101613 (2020)

20. Rashid, J., et al.: Phishing detection using machine learning technique. In: First International Conference of Smart Systems and Emerging Technologies. IEEE (2020)
21. Geyik, B., Erensoy, K., Kocyigit, E.: Detection of phishing websites from URLs by using classification techniques on WEKA. In: Proceedings of the 6th International Conference on Inventive Computation Technologies. IEEE (2021)
22. Korkmaz, M., Sahingoz, O.K., Diri, B.: Detection of phishing websites by using machine learning-based URL analysis. In: Proceedings of the 11th International Conference on Computing, Communication and Networking Technologies. IEEE (2020)
23. Patil, V., et al.: Detection and prevention of phishing websites using machine learning approach. In: Fourth International Conference on Computing Communication Control and Automation. IEEE (2018)
24. Ahammad, S.H., et al.: Phishing URL detection using machine learning methods. Adv. Eng. Softw. **173**, 103288 (2022)
25. Hannousse, A., Yahiouche, S.: Towards benchmark datasets for machine learning based website phishing detection: an experimental study. Eng. Appl. Artif. Intell. **104**, 104347 (2021)
26. Chopra, D. and Khurana, R. (2023) *Introduction to Machine Learning with Python*: Bentham Science Publishers
27. Jarrett, C.: Categorical features in XGBoost without manual encoding (2023) https://developer.nvidia.com/blog/categorical-features-in-xgboost-without-manual-encoding/#:~:text=Now%2C%20XGBoost%201.7%20includes%20an,without%20having%20to%20manually%20encode
28. Malakouti, S.M., Menhaj, M.B., Suratgar, A.A.: The usage of 10-fold cross-validation and grid search to enhance ML methods performance in solar farm power generation prediction. Cleaner Eng. Technol. **15**, 100664 (2023)
29. Sahingoz, O.K., Buber, E., Demir, O., Diri, B.: Machine learning based phishing detection from URLs. Expert Syst. Appl. **117**, 345–357 (2019)

Online Gambling in the Rural Global South: Probably the Next Major Silent Killer

Willem van Eekelen$^{(\boxtimes)}$ ⓘ

Independent Evaluator, Codsall, UK
`willem.vaneekelen@gmail.com`

Abstract. This paper combines the author's evaluative datasets with nascent literature on rural gambling, to come to a narrative about a worsening epidemic of online gambling disorders in the rural Global South. The paper posits that the rapid spread of mobile money enabled gambling companies to penetrate ever deeper into regions they could not previously access. This has generated a significant and growing flow of money from rural regions to distant companies, and a wave of gambling disorders among rural men. There is evidence that online gambling is increasingly affecting rural women as well. The offer is enormous and it is available 24/7, from the privacy of one's home. Regulation is rudimentary and restrictions are ineffective. In most rural regions harm-reduction support is not available. Online gambling is likely to be the next major silent killer in the rural Global South.

Keywords: online gambling · online betting · problem gambling · gambling disorders · harmful behaviour · mobile money · rural · AI · Grand Challenges · Global South

1 Introduction and Methodology

When the Covid-19 pandemic halted travel, my work as an evaluator of international development efforts moved from programme sites to my kitchen table. This affected the quality of my data and analysis so I paused my work until travel had resumed.

Fighting boredom, I looked at the evaluative data I had gathered since 2010, from across some 35 countries in the Global South. My aim was to find patterns across these data sets. My hope was that such patterns might contribute to an overall understanding of the complex and tenacious Grand Challenges the world is facing today, in a way that individual evaluations and conventional research are unlikely to do because of their need for pre-defined investigative boundaries.

It was a rewarding exercise. It led to a late-in-life PhD on 'rural development' writ large, and to a book titled *ICT and rural development in the Global South* [1].

This paper is yet another product of that exercise, and an updated and reworked version of chapter 11 of the book I just mentioned. I wrote this paper for the September

---

W. van Eekelen—Independent Evaluator

© IFIP International Federation for Information Processing 2024
Published by Springer Nature Switzerland AG 2024
R. M. Davison and D. Kreps (Eds.): HCC 2024, IFIP AICT 719, pp. 86–97, 2024.
https://doi.org/10.1007/978-3-031-67535-5_8

2024 HCC16 conference on "Humans, Technological Innovations and Artificial Intelligence: Opportunities and Consequences", under the theme of "Global or local: how are the lived experiences of marginalised and dispossessed citizens in developing countries affected by new technological innovations?".

It is not a conventional academic paper, and not even a summary of insights gained from gambling-related evaluations (in fact I have never evaluated anything that explicitly focused on gambling). Instead, this paper is based on gambling-related data and observations I more or less accidentally gathered since 2010, which I then compared and contrasted with the nascent literature on rural gambling to come to a narrative on what appears to be a worsening epidemic of online gambling disorders.

2   Gambling Problems Used to be Largely Urban Problems, but this is Changing

In a discussion paper prepared for the World Health Organization (WHO), Max Abbott says that "the gambling-related burden of harm appears to be of similar magnitude to harm attributed to major depressive disorder and alcohol misuse and dependence". He argues that the harm caused by gambling "is substantially higher than harm attributed to drug dependence disorder" and that "the burden is primarily due to financial impacts, damage to relationships and health, emotional/psychological distress and adverse impacts on work and education. This burden is disproportionately carried by disadvantaged and marginalised population sectors and contributes to health and social disparities" [2, p. 1, 2, 3]. The evidence base for these statements is slim as Abbott's paper was based on research in New Zealand and Australia only. In my experience the damage caused by alcohol still far exceeds the damage caused by gambling throughout the Global South. Moreover, gambling disorders as a large-scale problem have only just arrived in many rural regions.

When I first saw a concentration of problem gamblers in the Global South, the gambling industry was still almost entirely urban-based and urban-focused. It was in 1990, in Agadez, a Nigerien town at the edge of the Sahara Desert. Mostly young men (no women)[1] gathered in and around a gambling hall that was profitable enough for the French owner to be able to live in a villa that featured the town's only swimming pool. It was dark and gloomy inside, with lines of machines, and I remember this place so well because of the desperation on the faces of the men as they walked out. It reminded me of my years as a volunteer in a homeless shelter in Rotterdam, where roughly a fifth of the guests were homeless because of a gambling disorder.

The gambling hall in Agadez had a grave impact on problem gamblers, but the number of people affected was limited as this was a physical place, in the centre of town, and it was too long a walk for rural men to cause impulse gambling. For obvious reasons of customer density, gambling halls and betting shops commonly choose urban locations. There are rows of betting shops in Kampala, for example, but I have never seen one in any of Uganda's small villages (except for those that are situated along a main road). The link between gambling and the proximity of gambling opportunities is well-evidenced, and means that the likelihood of lifetime gambling (i.e., of people ever having gambled), of impulse gambling and of building a gambling disorder has always been far lower in rural regions than in urban ones [7–9].

Indeed, as an evaluator and until a few years ago, I had only very occasionally come across rural problem gamblers. A few men had used microfinance facilities to borrow money, or had forced their wives to borrow money, and had gambled it away. A blind woman in Malawi had received training and other support and, as a consequence, was able to grow crops in her garden – but she did not benefit from the harvest as her husband sold it and used it to feed his drinking and gambling habits. I remember these specific tragedies because they were unusual: in most rural programmes, gambling did not come up as an impediment for anything. However, this is changing, and over the past few years gambling has become one of the themes that regularly appears when I ask rural people about their sources of income and expenditure.

---

[1] Studies that focus on Western countries show roughly comparable gambling participation rates of men and women. For example, Simone McCarthy and her colleagues [3] refer to four studies that came to this conclusion. However, Africa- and Asia-focused studies [4, 5, 6] conclude that men gamble more than women; that problem gambling is more common among men than among women; and that women are often opposed to the gambling habits of their husbands and other male household members.

Research has not kept pace with this new phenomenon. Gambling in both Asia and Africa is under-researched[2] and, because rural gambling has traditionally been far less of a problem than urban gambling, rural gambling is rarely researched at all. I found only two pieces of research that covered rural gambling, anywhere in the Global South, and both confirmed the gap between urban and rural gambling prevalence. First, a systematic review of gambling research in Africa [4] found that lifetime gambling was prevalent in well over half (57%–73%) of youth (or sometimes of young men only, depending on the research) – but the only *rural* study in the review, conducted in Malawi, found only 16% prevalence of lifetime gambling [12]. Second, research among 150 rural and 150 peri-urban individuals in South Africa's KwaZulu-Natal (both groups with equal numbers of men and women) found that 75% of the people who had never gambled were rural people; that the vast majority of the people at risk were urban people (68% of people at low risk and 90% of people at moderate risk were urban); and that *all* the people assessed to be problem gamblers were urban people [13].

When rural people do gamble, they have traditionally been more likely than urban people to do this in card games, with dice, through coin spinning and by betting on animal fights (generally cockfights, but in Indonesia I saw a lively trade in betta, a fighting fish used in gambling). This is generally *personal* gambling: it is done in social groups, without a company that is taking a cut (though there might be people who receive payments, such as formal authorities if gambling is illegal, or the organiser or host).[3] Some research found that such forms of gambling can still be damaging for chronic losers [15] and that gambling debts may be a reason to migrate in order to escape from creditors [6]. Other research found no measurable effects on the long-term distribution of wealth, [16] or concluded that gambling was largely harmless as winners were expected to share their gains, as "a big win [...] is socialised cash. [The winners] can expect to receive loan requests. Quietly stashing away money is seen as stingy, and while the cash may be hidden from the living, it places the individual in danger of supernatural sanctions." [14, p. 112]. Either way, for a community as a whole, personal gambling is far less damaging than commercial, *impersonal* gambling, where the bettor plays against a slot machine or similarly faceless online betting instrument. This is because betting without a third party taking a cut is, in the long run and on average, budget-neutral for the individual gamblers (provided it does not require skills), and the money does not leave the community. I came across one piece of research that concluded that such gambling could even have a *positive* effect on the local economy, as "the intense desire to play

---

[2] In Asia, gambling research tends to focus on specific locations (e.g., Hong Kong, Macao and Singapore). For example, a 2017 review of gambling research in India concludes that "To the best of our knowledge, there have only been three studies of gambling from India" [10, p. 41]. A passing reference to this statement, made in 2020 by the same lead author, suggests that nothing new had appeared in the last few years [11]. In Sub-Saharan Africa, a 2021 systematic review of publications on youth gambling (and note that most gambling research focuses on youth) found only 13 publications that reported on data-driven research in peer-reviewed journals in all of Sub-Saharan Africa [4].

[3] Organised group gambling does have systematic 'leakage'. In cockfights in Flores in Indonesia, for example, "winners double their money, [minus] ten per cent for the house, five per cent for the landowner, and five per cent for security – army and police" [14, p. 109].

cards for money [...] is a powerful force motivating people to engage in cash-earning activities" [15, p. 204].

But rural habits have always evolved through exposure to urban and foreign practices, and this is also true in relation to gambling. In the past, card game gambling either introduced gambling to rural regions [16] or replaced indigenous forms of gambling [14]. In more recent years rural regions have seen an advent of impersonal betting. This started with scratch cards. At least in Ghana and more recently, Chinese slot machines have become sufficiently affordable to present an attractive source of extra income for rural shop owners [17].

Such scratch cards and slot machines bring gambling within reach of many rural people – but impulse gambling is limited to people who are visiting a shop that offers gambling options, or who live relatively close to such a shop. The more threatening trend started with radio stations capitalising on the spread of mobile phones by opening betting lines that entice people to send charged text messages to answer easy questions, with the chance of winning money. These gambling options still exist, but in recent years they came to operate in parallel to betting via smartphones, and betting companies are now able to reach into people's own homes, also increasingly deep into rural regions, 24/7. In these rural regions, such companies find an audience that is large and, when it comes to gambling, naïve and inexperienced.

3 Mobile Money is Facilitating the Penetration of Impersonal Gambling Products into Rural Regions

One of the negative side effects of financial ICT innovations is that they may cause financial despair. M-Pesa is a case in point. M-Pesa became the poster child for rapid financial inclusion and was much celebrated and promoted, especially after the publication of a paper by Tavneet Suri and William Jack, published in *Science*, which attributed large-scale poverty reduction to M-Pesa's mobile money [18]. This claim was often used by the World Bank and other stakeholders working in the field of financial inclusion.[e.g. 19] The problems mobile money proponents did mention were put down to mere technical glitches. Failed transactions or delayed confirmation messages were attributed to congestion during peak texting times, and an inability to pay out on the side of rural M-Pesa agents was the consequence of cash flow constraints, for example [20]. However, Suri and Jack's underpinning assumptions were refuted in 2019 [21]. Since then, more negative research has appeared, often about unregulated mobile money providers causing indebtedness that reinforces poverty. Over 2.7 million Kenyans are blacklisted after digital loan defaults, [22] for example, and the April 2019 *FinAccess Household Survey* of Kenya [23] presents a bleak picture of the effects of digital microfinance. As summarised by Sibel Kusimba:

This extensive survey of consumer finance and digital finance showed that in spite of the greater reach of finance and credit products into people's lives, their overall financial health had greatly declined since 2016. Only 20% of Kenyans – all income levels – were financially healthy, as measured by their ability to save for emergencies, plan, and meet their daily needs (down from 39% in 2016). Many had lost money to digital scams. Although inclusion in terms of accounts and usage of products has increased, people are not benefiting. It is hard to reconcile the celebratory narrative of M-Pesa with the consumer risks it soon created [24, p. 47].

This survey differentiated between rural and urban areas, and found the situation in rural areas to be far worse than in urban ones: only 14% of rural people were in good financial health, compared to 33% of urban people [23].

Part of this problem of indebtedness is caused by online and other forms of ICT-powered gambling. Gambling is within the realm of the 'recreational use' of ICT applications. Recreational use is the world's most common type of individual ICT use, but it and its risks and dangers are widely ignored by researchers within the field of rural development and by the aid industry alike [25]. This blind spot applies to online gambling as well. I have not heard of research on the parallels between the surge in online gambling and the equally sudden rise in pyramid and Ponzi schemes in former communist countries, in the early years after the end of the Cold War, for example [26, 27] Similarly, I have searched for but have not seen a single pre-2019 donor paper or academic publication that predicted that (initially donor-funded) mobile money products might create a problem of indebtedness by suddenly exposing an unprepared population to the heavily advertised prospect of a new way of making money through gambling. Online gambling was a 'Black Swan' – a phenomenon that was easy to explain after its existence had become obvious, but an unanticipated and suddenly life-changing phenomenon in its early stages. As was the case with the pyramid and Ponzi schemes in Eastern Europe, new forms of gambling, powered by the instant link between gambling options, mobile money wallets and microloans, is already causing major setbacks in the financial position of millions of people throughout the rural Global South. In absence of other forms of impersonal gambling, the spread is entirely caused by ICT-facilitated gambling products, and most such products are accessed through mobile phones and smartphones. Their rapid spread is bringing evermore rural regions within easy reach of the gambling industry.

If the trend continues then Abbott's statement about comparability of harm caused by gambling disorders and alcohol misuse may well become true in the very near future. This is because the new rural customer base is no match for online gambling companies that use artificial intelligence to set odds and change them in real time, segmentalize the market and personalise their offer and visuals. Their collective offer is enormous. I checked a Ugandan interface app for gamblers and found that it offers 11 companies to gamble through (Kagwirago, BetWay, BetPawa, EliteBet, BetLion, WorldStar, Aba-Bet, Betin, GrandVictoria, Next and Betcity), each of which has a wide range of betting options at any time. The industry is far larger still in Kenya.[4] The Kenyan (but partly Bulgarian-owned) SportPesa illustrates the size of the country's online betting business. It has signed sponsorships with the English Premier League football clubs Hull City, Arsenal, Southampton and Everton, and in the cases of Hull City and Everton the

---

[4] The "total financial turnover [of the top 12 Kenyan websites only] is estimated at 250 billion [Kenyan shillings, which, at the time, equated to some US$2.4 billion], more than the national recurrent budget". [28, p. 122] [The national recurrent budget is the part of the government budget that covers ongoing expenses (as opposed to capital expenses)].

sponsorships were the best-paying sponsor agreements in the clubs' histories [29, 30][5]. Even where online betting is illegal, such as in much of India, "the constantly increasing technological advancement and internet penetration, coupled with easy accessibility to and affordability of smartphones, [gives rise to] a growing concern that more people will start gambling online, and subsequently, more people will have gambling-related problems" [11, p. 2].

This online gambling trend is dangerous. Online gambling is new and it will take time for the negative effects to become sufficiently painful, widespread and visible to create awareness of its dangers. For now, gambling is widely seen as a legitimate activity, [4] and this is reinforced by those sponsorships in the English Premier League, which is by far the world's most-watched and betted-on football league.

Some people – and mostly adult men – still prefer to place their bets in betting shops or social venues such as tea stalls, even if they have online gambling accounts, because of the social dimension, [31, 32] and perhaps because betting companies recruit attractive women to staff their shops.[6] As a 20-year-old man from Bangladesh attests, such social settings have the drawback of peer pressure:

See Apu [Apu means sister and refers to Sharifa Sultana, the interviewer], all of us do not play. But some of our bettor friends will still call us by name and say, "of course you do not play, because you do not have money to play. Or you do not have any guts to play- oh Mamma's boy." This is ridiculously insulting and many of us even occasionally play to avoid such conversations [32, p. 11].

For many others, online gambling has the advantage of being discreet ("I love my privacy. This makes me gamble mostly online") [31, p. 70]. They gamble in isolation, and research in the arrangement of casino slot machines suggests that gambling in isolation impairs a gambler's control:

Probable pathological gamblers preferred the cubicle or the isolated area arrangement. They did not like the against-the-wall and counter layouts because these were considered too visible and too susceptible to the distractions of the surroundings. Despite this preference, pathological gamblers unanimously believed that the isolated area promoted impaired control over their gambling habits [33, p. 144].

Online gambling pulls in people who are too young to be allowed entry into betting shops, even after considering that age restrictions are widely ignored in such shops [34]. 'Safety measures' such as the requirement to click a button saying "I confirm that I am over 25 years old" do not pose a meaningful obstacle, and mobile money accounts can easily be borrowed from friends and older siblings. Online gambling draws in women, too. There were no women in that gambling hall in Agadez, back in 1990, or in many of Africa's and Asia's betting shops today, and in my evaluative work I have not encountered cases where livelihood interventions failed because of a woman's gambling. However, cultural norms matter less in the privacy of one's home and there is early (and as yet

---

[5] Everton has discontinued the agreement prematurely, following mounting criticism about SportPesa's role in the 'gamblification' of football.

[6] I base this on my own observations and an interview with a young East African woman, with whom I discussed livelihoods options, and who matter-of-factly mentioned that there was a group of opportunities, such as running a street-side tea shop or working in a betting shop, that she did not qualify for as she was insufficiently attractive.

scant) evidence that the "numbers [of young women gambling] have recently started to increase for online betting" [35, p. 36].

Importantly, online betting is *always* available, easy to access and directly linked to mobile money wallets. These mobile wallets advance financial inclusion and have many useful roles to play, but for people who engage in online betting, harm may outweigh these benefits. Research among a little over a thousand (1,040) digital borrowers in Kenya, of whom some 30% (304) were 'digital bettors', concluded that "bettors were shown to be more likely than non-bettors to be financially distressed, engage in welfare undermining coping strategies, and have inferior welfare outcomes" [36, p. 10]. Causality goes both ways: gambling can *cause* financial distress, and gambling can be a form of negative coping behaviour for people *in* financial distress.

In many ways the digital divide between urbanites and rural people reinforces existing inequalities, and narrowing it would bring ample benefits. However, narrowing this divide would also mean an expansion of the problems of online gambling into ever-more remote rural regions.

4 Strong Measures Could Reduce the Risk of Online Gambling Turning into a Major Silent Killer in the Rural Global South

A growing betting industry does not have to be a problem. In a paper I quoted before, Max Abbott (at the time the Chairman of the International Gambling Think Tank) said that:

There is little doubt that greater gambling availability has led to increased consumption and increased problems in many parts of the world. However, in both expanding and maturing markets [...] problem gambling rates can decline, sometimes markedly, rather than increase [2, p. 4].

Realising a reduction of gambling-caused problems in expanding gambling markets requires, at least, a change in the image of gambling, enforced restrictions, and support systems for problem gamblers. Each of these three requirements is difficult to achieve at scale.

It is Possible but Hard to Change the Image of Harmful Behaviour. Graphic health warnings have proven effective in discouraging non-smokers from starting to smoke [37] (though the extent of the effectiveness depends on, among other things, the size, frame and type of pictures, and even on the colour of the accompanying text) [38] However, at the moment there are few comparably stern warnings in relation to gambling, while there is a flood of encouraging advertisements – in-country or, where this is illegal, from offshore companies. Moreover, even highly stern and visible messages are unlikely to provide sufficient counterweight to advertising such as in the Premier League. As long as so much of the betting is focused on an industry that promotes it, it will remain within the realm of the acceptable and welcome but partial rules such as refraining from front-of-shirt advertising (which will soon come into effect in the Premier League)[7] are unlikely to change this. At central levels, changing the image of gambling would require,

---

[7] In April 2023, the Premier League clubs agreed to stop featuring gambling adverts on the front of their shirts, as of the 2026-27 football season. [39]

at least, a wholesale ban on advertising and a concerted efforts to convey the dangers of gambling. In many countries this is unlikely to happen, if only because any perception of duty of care government's may have in this field is unlikely to withstand the pressure of the powerful gambling lobby, which "uses its enormous political power to accelerate deregulation, expand its remit, and resist public health reforms" [40, p. 2, 41]. At local levels there needs to be better visibility of the harm caused by problem gambling, up to and including suicides, and this requires strong and sustained local efforts to break the taboo on gambling debts and their consequences.

It is Just as Hard to Enforce Gambling Bans or Restrictions. Outright bans are unlikely to address the problem of problem gambling. Online gambling is illegal in all but three states in India (Goa, Daman and Sikkim), and other forms of gambling are highly restricted, yet annual gambling revenues in and from India amount to tens of billions of US dollars (though admittedly nobody knows how much it really is[8] or how much more it might have been without these bans and restrictions). Regulation could conceivably be more effective. Restrictions on promotions (e.g., first bet free, every tenth bet free) on radio, mobile phones and smartphones could reduce the incentive to start or continue gambling. Enforced betting breaks and pop-up warnings could potentially have a similar effect, and so could limits on access time, bet size, numbers of bets placed, and speed of play. In the rural Global South, such restrictions are not yet commonly used. Rather the contrary: companies use artificial intelligence to tailor their promotions to the vulnerabilities of individual gamblers, and online companies introduced high-speed betting options such as three-minute virtual football games so that gamblers do not have to wait 90 min of play time before knowing the result.

Rural Problem Gamblers in the Global South Need Support But are Difficult to Identify and Reach. Like drug use disorders, gambling disorders are, for now, a mostly urban problem, and harm reduction support for people with such disorders has yet to be developed in rural regions.[9] However, internet-delivered interventions for problem gambling [43] (and gaming) [44] are promising. Such interventions align with the finding that problem gambling is best dealt with in a process of self-directed change and therapy. They could potentially reach people in remote rural regions, though they cannot reach large groups without major investments because of the importance of elements such as motivational interviews as part of therapy. Perhaps more importantly, such approaches require people and the communities they live in to recognise that gambling may be a problem that requires change and therapy. For such recognition to become

---

[8] The only estimate I know of is more than a decade old (from 2013): US$42 billion. [42]

[9] I base this last statement on work I did in 2013–2016, for a programme titled 'Asia Action on Harm Reduction'. This programme focused on organisations supporting (and often owned and managed by) drug users. These organisations, and the organisations in their wider networks, all focused almost exclusively on urban centres (and occasionally key transport hubs). The statement makes two assumptions. First, Asia Action was focused on drug users, not problem gamblers, and I assume a parallel because both are predominantly urban phenomena in which harm reduction work takes the shape of small-scale interventions. Second, my work for Asia Action took place in Asian countries only (Cambodia, China, India, Indonesia, Malaysia and Vietnam) and I merely assume that the situation is similar in Africa on the basis of never having come across support facilities that did exist in rural Africa but not in rural Asia.

widespread rural online gambling in the Global South is likely to first have to grow into a problem of such enormous proportions that even the privacy of gamblers' homes no longer keeps it silent and invisible.

References

1. Eekelen, W. van.: ICT and rural development in the Global South, Rethinking Development Series, Routledge (2024)
2. Abbott, M.: "The epidemiology and impact of gambling disorder and other gambling-related harm", Discussion paper for the 2017 WHO Forum on alcohol, drugs and addictive behaviours, World Health Organization (2017)
3. McCarthy, S., Thomas, S.L., Bellringer, M.E., Cassidy, R.: Women and gambling-related harm: a narrative literature review and implications for research, policy, and practice. Harm Reduct. J. **16**(18), 11 (2019)

4. Bitanihirwe, B.K.Y., Ssawanyana, D.: Gambling patterns and problem gambling among youth in Sub-Saharan Africa: a systematic review. J. Gambl. Stud. **37**(3), 723–745 (2021)
5. George, S., et al.: "A cross-sectional study of problem gambling and its correlates among college students in South India". BJPsych (i.e., British J. Psych.), **2**(3), 199–203 (2016)
6. Nooteboom, G.: Forgotten people: poverty, risk and social security in Indonesia; the case of the Madurese, Brill (2015)
7. Tagoe, V.N.K., Yendork, J.S., Asante, K.O.: Gambling among youth in contemporary Ghana: understanding, initiation and perceived benefits. Africa Today. **64**(3), 52–69 (2018)
8. Pearce, J., Mason, K., Hiscock, R., Day, P.: A national study of neighbourhood access to gambling opportunities and individual gambling behaviour. J. Epidemiol. Commun. Health **62**(10), 862–868 (2008)
9. Sévigny, S., Ladouceur, R., Jacques, C., Cantinotti, M.: Links between casino proximity and gambling participation, expenditure, and pathology. Psychol. Addict. Behav. **22**(2), 295–301 (2008)
10. George, S., Velleman, R., Nadkarni, A.: Gambling in India: past, present and future. Asian J. Psych. **26**, 39–43 (2017)
11. George, S., Fenn, J., Robonderdeep, K.: An overview of gambling in India. Global J. Med., Pharm. Biomed. Update. **15**(4), 1–4 (2020)
12. Muchimba, M., et al.: Behavioral disinhibition and sexual risk behavior among adolescents and young adults in Malawi. PLoS ONE **8**(9), 6 (2013)
13. Dellis, A., Spurrett, D., Hofmeyr, A., Sharp, C., Ross, D.: Gambling participation and problem gambling severity among rural and peri-urban poor South African adults in KwaZulu-Natal. J. Gambl. Stud. **29**(3), 417–433 (2013)
14. Curnow, J.: Gambling in Flores, Indonesia: not such a risky business. Australian J. Anthropol. **23**, 101–116 (2012)
15. Grossman, L.S.: Peasants, subsistence ecology, and development in the highlands of Papua New Guinea, Princeton University Press (1984)
16. Hayano, D.M.: Like eating money: card gambling in a Papua New Guinea highlands village. J. Gambl. Behav. **5**, 231–245 (1989)
17. Hayk, A., Sailer, U.: Cosmopolitan encounters provoke a change in habits: how Chinese slot machines affect rural life in Ghana. Geoforum **111**, 39–47 (2020)
18. Suri, T., Jack, W.: The long-run poverty and gender impacts of mobile money. Science **354**(6317), 1288–1292 (2016)
19. Dawson, S.: Why does M-PESA lift Kenyans out of poverty? Consultative Group to Assist the Poor (2018)
20. Morawczynski, O., Pickens, M.: "Poor people using mobile financial services: observations on customer usage and impact from M-PESA", CGAP Brief, Consultative Group to Assist the Poor, 4 pages (2009)
21. Bateman, M., Duvendack, M., Loubere, N.: Is fin-tech the new panacea for poverty alleviation and local development? Contesting Suri and Jack's M-Pesa findings published in Science. Rev. African Political Econ. **41**(161), 480–495 (2019)
22. Gitonga, S.: Millions of Kenyans blacklisted by CRB over digital loan defaults, Business Today (2019)
23. CBK, KNBS and FSD Kenya (April 2019) FinAccess household survey; access, usage, quality, impact, Central Bank of Kenya, Kenya National Bureau of Statistics, and Financial Sector Deepening Kenya (2019)
24. Kusimba, S.: Reimagining money; Kenya in the digital finance revolution, Stanford University Press (2021)
25. Arora, P.: The next billion users: digital life beyond the West, Harvard University Press (2019)
26. Schiffauer, L.: "Dangerous speculation; the appeal of pyramid schemes in rural Siberia", Focaal –. J. Global Histor. Anthropol. **81**, 58–71 (2018)

27. Verdery, K.: "Faith, hope and Caritas in the land of the pyramids, Romania, 1990–1994", chapter 7 in Verdery, K., What was socialism, and what comes next?, Princeton University Press (1996)

28. Amutabi, M.N.: Gambling addiction and threat to development in Kenya: assessing the risks and problems of gamblers in changing society. J. African Interdiscip. Stud. **2**(2), 121–133 (2018)

29. Conneller, P.: EPL team Everton aborts SportPesa sponsorship deal two years early, Casino.org (2020)

30. SportPesa (undated) SportPesa joins up with the tigers!, SportPesa

31. Adebisi, T., Alabi, O., Arisukwu, O., Asamu, F.: Gambling in transition: assessing youth narratives of gambling in Nigeria. J. Gambl. Stud. **37**(1), 59–82 (2021)

32. Sultana, S., Mozumber, M.H., Ahmed, S.I.: "Chasing luck: data-driven prediction, faith, hunch, and cultural norms in rural betting practices". In: CHI Conference on Human Factors in Computing Systems (CHI '21), Yokohama, Japan (2021)

33. Ladouceur, R., Jacques, C., Sévigny, S., Cantinotti, M.: Impact of the format, arrangement and availability of electronic gaming machines outside casinos on gambling. Int. Gambl. Stud. **5**(2), 139–154 (2005)

34. BBC News Africa Gamblers like me: the dark side of sports betting, BBC (2019)

35. Mustapha, S.A., Enilolobo, O.S.: Effects of gambling on the welfare of Nigerian youths: a case study of Lagos State. J. Gambl. Issues **43**, 29–44 (2019)

36. Chamboko, R., Guvuriro, S.: The role of betting on digital credit repayment, coping mechanisms and welfare outcomes: evidence from Kenya. Int. J. Fin. Stud. **9**(10), 12 (2021)

37. Drovandi, A., Teague, P.A., Glass, B., Malau-Aduli, B.: "A systematic review of the perceptions of adolescents on graphic health warnings and plain packaging of cigarettes". Syst. Rev. **8**(25) (2019)

38. Gantiva, C., Mejia-Orjuela, C.: "Yellow warnings labels on top are more effective to discourage smoking initiation: an experimental online study". Am. J. Health Educ. 1–6 (2024)

39. SIG: Premier League confirms gambling sponsorship ban, Sport Industry Group (2023)

40. Markham, F., Young, M.: 'Big Gambling': The rise of the global industry-state gambling complex. Addict. Res. Theory **23**(1), 1–4 (2024)

41. Mandolesi, G., Pelligra, V., Rolando, S.: "The gambling industry's corporate structure in a partially liberalised market", a chapter in Nikkinen, J., Marionneau, V., Egerer, M. (Editors) The global gambling industry: structures, tactics, and networks of impact, Springer (2022)

42. George, S., Velleman, R., Weobong, B.: Should gambling be legalised in India? Indian J. Psychol. Med. **43**(2), 163–167 (2020)

43. Moreira, D., Dias, P., Azeredo, A., Rodrigues, A., Leite, Â.: A systematic review on intervention treatment in pathological gambling. International Journal of Environmental Research and Public Health (IJERPH) **21**(3), 18 (2024)

44. Park, J.J.: The development, feasibility, and effectiveness of internet-delivered interventions for problem gaming, PhD thesis, University of Auckland (2024)

The Digital Transformation of Microbusinesses in Indonesia: Dichotomous Effects and Consequences

Mahendrawathi ER[1]([✉]) [iD], Robert M. Davison[2] [iD], Safirotu Khoir[3] [iD],
and Rahma Fauzia[1] [iD]

[1] Institut Teknologi Sepuluh Nopember, Surabaya, Indonesia
mahendrawathi.er@its.ac.id
[2] City University of Hong Kong, Kowloon, Hong Kong
isrobert@cityu.edu.hk
[3] Universitas Gadjah Mada, Yogyakarta, Indonesia
safirotu@ugm.ac.id

Abstract. Digital transformation is widely lauded as being beneficial for organizations. We undertake a qualitative exploration into how three microbusinesses in Indonesia are affected by Digital Transformation. We focus on the lived experiences of the microbusiness owner-managers. We find that the success of a digital transformation initiative depends on the owner-managers having both a growth and a technology mindset while observing that the availability of support from friends and relatives can be crucial to success.

Keywords: Digital Transformation · microbusiness · warung · small retail · sociotechnical systems

1 Introduction

Digital transformation (DT) is a phenomenon of our time. The allure of DT is powerful, and the literature tends to paint a rosy picture for larger and smaller organizations, including Small and Medium Enterprises that employ fewer than 250 people and Microbusinesses (MBs) that employ fewer than 10 people (OECD, 2023). For instance, it is acknowledged that DT can reduce costs and save time and resources. This is also true for small and medium-sized enterprises (SMEs), even though they have limited market size, bargaining power and fewer internal capabilities to deal with complex business environments (Vide et al., 2022). On the other hand, DT attempts are not always successful, which is perhaps not so surprising given the tendency of people to resist the changes associated with such transformational events. However, few scholars or pundits would argue that the idea of DT is itself a bad thing. Instead, the objective of redefining an organization's value proposition with digital technology is widely appreciated. But is DT really an activity that is globally appropriate?

© IFIP International Federation for Information Processing 2024
Published by Springer Nature Switzerland AG 2024
R. M. Davison and D. Kreps (Eds.): HCC 2024, IFIP AICT 719, pp. 98–109, 2024.
https://doi.org/10.1007/978-3-031-67535-5_9

As a developing country, the number of SMEs in Indonesia is significant. 99% of organizations in Indonesia are SMEs or Microbusinesses (MBs), employing fewer than 10 people (Indonesia Investment, 2022). These 64.2 million SMEs (Harian Surabaya, 2024) contribute more than 61% of the gross domestic product (GDP), employ 97% of the labour force and attract more than half of the total investment in Indonesia. Meanwhile, several initiatives to support digitalization for SMEs have been undertaken (e.g. the 2018 Making Indonesia 4.0 Roadmap, the 2019 E-Commerce Roadmap, and the 2020 Go Digital Vision) (Hermawan, 2022). However, we need to observe more deeply how such initiatives have led to success, particularly for MBs.

Some initiatives have been conducted to support the DT of SMEs or MBs. For example, Yogyakarta province created the SiBakul app as part of its commitment to support SMEs and MBs. SiBakul is a digital platform for the guidance and development of cooperatives and SMEs, as well as marketing facilitation of SME products officially managed by the local authority. Further, in Surabaya, a commercial telecommunications company (Telkomsel) dedicated their Corporate Social Responsibilities in the last three years to support SMEs by creating a Digital Creative Entrepreneurs (DCE) programme (Harian Surabaya, January 24, 2024).

In this paper, we explore how the owner-managers of warungs, a unique form of MB in Indonesia, grapple with the technology needed for a DT of their businesses. A warung, is a small family-operated business that can serve various purposes like dining, coffee shops, or local convenience stores. In this discussion, we focus on the convenience store type of warung, often referred to as a mom-and-pop store. Our research question is: how are the lived experiences of warung owner-managers affected by DT? Following this introduction, we review the literature on DT, with a strong focus on MBs, and on warungs. We then introduce our research method, a case study, and explain how we collected and analysed interview data. We present three cases of warungs and their owner-managers, thematically organized around the central notion of how DT affects the warung. We discuss our findings in the context of the DT and MB literature, as well as the broader aspect of how information systems are deployed in developing countries. Finally, we conclude with an assessment of the practical and theoretical contributions, the limitations, and future research directions.

2 Literature Review

Digital transformation (DT) is a prominent fixture in the organizational change landscape. DT is generally accepted as encompassing activities that leverage technology, including websites, social media, mobile, data analytics, cloud, and e-commerce (Shirish et al., 2023) in order to redefine the value proposition of an organization, including the emergence of a new organizational identity (Wessel et al., 2021). Although much of the literature on DT focuses on large organizations in the developed countries of the global North, there is an increasing interest in the experiences of smaller organizations in the Global South. For instance, it has been observed that digital technologies have helped small firms transform their business models in the wake of the COVID-19 pandemic (Priyono et al., 2020). Furthermore, these same digital technologies help smaller firms build resilience during a crisis (Khurana et al., 2022). However, the DT of an

SME requires owners to have heightened awareness and learning so that they can sense, seize, and transform at the individual, organizational and ecosystem levels. Owners must, therefore, search for information that pertains to the digital options available.

The unique characteristics of MBs have been addressed in many streams of literature including IT transformation, e-business, and DT. For instance, it is noted that MBs are agile, dynamic, interactive, less formal, and bureaucratic than larger organizations (Bai et al., 2021). This 'smallness' can be a positive factor because MBs can quickly initiate DT (Mandviwalla & Flanagan, 2021). However, institutional smallness also corresponds to limited IT infrastructure, personnel, budget, and expertise (Mandviwalla & Flanagan, 2021). These limitations can be overcome by informally bringing external people into the MB, including family and friends. For instance, Mkansi (2022) found that successful MB e-business adoption depends on cooperation among both people and ecosystems. An interpersonal ecosystem provides many resources for MBs, including people who can act as financial sponsors, cross-pollinators of knowledge, and complementary staff all of whom can provide MB owners with the space to focus on business activities. It is common in MBs for the owner to depend on family members especially for marketing and staffing. MBs can also leverage external ecosystems, e.g. third party DT consultants or solution providers, as they seek to manage the DT process. Mandviwalla and Flanagan (2021) observe that MB owners often see the promise of DT and can quickly move from thinking about and specifying a problem to describing a desirable solution.

Apart from the high levels of interest in a DT solution to a business problem, MB owners who are DT aware are likely to take actions that will enhance the further development and sustainability of the MBs themselves. Shirish et al. (2023) created a three-tier (high, medium or low) classification of the 'growth mindset' of MB owners. A similar classification arrangement, based on attitude towards technology (proactive, reactive or passive), helps us understand how MB owners are likely to approach DT in practice.

In the Indonesian context, a particular kind of MB is locally known as a warung. The warung constitutes part of the intangible heritage of the Indonesian culture (Isnurhadi, 2022). A warung is typically owned and operated by individuals, often members of a single family. It generally functions as a small-scale general store that sells goods and services appropriate to and valuable in the local community. In this context, warungs "are not businesses, but a part of the local community. People visit warungs not just to shop, but often also to catch up and chat with neighbors" (Isnurhadi, 2022).

Although warungs play an important role in the local Indonesian community, they face many challenges, not least from the rapidly spreading small-scale chain supermarkets. The supply chain for a warung is often complex and messy, with numerous inefficiencies. Warungs may have as many as a dozen sourcing channels, with high levels of price variation and credit accessible only to the larger warungs (Behera, 2020). Warungs are thus at a significant disadvantage as they must also compete with chain minimarts with their own distribution centers. Nevertheless, the broad reach of warungs across the entire Indonesian archipelago has attracted the attention of logistics service providers who hope to expand their business into this niche. Warungs that have already undertaken DT are connected with digital ecosystems that provide accounting and bookkeeping, B2B marketplace, and lending solutions. Digitally transformed warungs can also offer Payment Point Online Banking (PPOB) services to their customers. PPOB is a real-time service undertaken in collaboration with online banking so that data reconciliation occurs quickly and accurately. Some of the most common PPOB services are bill payments, e.g., electricity, water, telecommunications, and data packet purchases.

Sampoerna Retail Community (SRC) is an example of a warung ecosystem provider that aims to provide a digitally supported solution for MBs to source supplies. Sampoerna Retail Community (SRC) started in 2008 in the format of mentoring and education to provide new insights regarding the management of the traditional retail business. Initially, 57 warungs in Medan joined to get business assistance from the SRC team. The initiative yielded positive results and the shops that were accompanied managed to grow their business. After that, SRC was set to become an official program to develop MBs. The number of warungs supported increased from 57 stores to 4,000 stores in 2009. The program concept then developed into sustainable business assistance, including coaching on store neatness, arrangement of goods and layout, marketing strategies, and financial management. They also provide incentives for warung partners that accomplished certain milestones. Warungs that are members of SRC are also educated to contribute to the economy and social environment where they are located. By 2018, the number of retailers in SRC passed 100,000, and today, the number has further expanded to 225,000+ grocery stores spread throughout Indonesia. There are 6,900 SRC partner associations, called Paguyuban, that actively share knowledge and experience, aiming to increase the competitiveness of grocery stores and make a positive contribution to advancing MBs (Mukharomah et al., 2023).

Along with technology development, the SRC program was developed into the digital world with the launch of the AYO SRC applications in May 2019. These applications are an innovation to facilitate access for SRC members to share business knowledge, get information about Sampoerna's SME coaching, and facilitate the store management process (Yasa, 2019). The AYO SRC applications, including AyoToko, Ayokasir and PojokBayar, are used by SRC partners to carry out their tasks in managing warungs. Each of these applications has different functions, but they are all integrated with each other. In AyoToko, warung owners can order goods, get promotional information, and goods catalogs and review their profit and capital. PojokBayar is an application to buy credit and e-money such as OVO, GoPay, data packages, or electricity tokens. AyoKasir is an application used to facilitate the sale of goods. In a warung, products are usually not barcoded or price-stamped so the owner will have to remember the price. When a customer buys a product, the owner usually does not record the sales transaction. The AyoKasir application includes barcode scanning and sales recording. The warung owner can scan the products, and the application will calculate the total purchase. In this application, partners can record all the products they have in their warung.

SRC also provides gamification in AyoToko, in which users need to undertake various tasks, called missions, to obtain reward points. The mission includes such activities as filling out questionnaires to self-assess the condition of the warung, or simply organizing and maintaining the cleanliness of their warung. The mission can relate to promotional activities such as arranging certain products such as mineral water from a certain brand partner. Another mission is transacting at PojokBayar for the purchase of credit or tokens and registering with the BJPS (Social Health Insurance Administration Body). Reward points obtained from completing the mission can then be exchanged for products or discounts. This way, SRC provides incentives for warung partners to improve the overall structure and quality of their warung.

3 Method

Since our research asked a broad question about how the lived experiences of warung owner-managers are affected by DT, we conducted an exploratory case study with three warung owners. We employed socio-technical systems (STS) theory to lay the foundation for data collection and analysis. Based on the literature, we identify attributes of the STS components, adapting them to reflect the specific warung case. We then developed an interview protocol with semi-structured questions to guide the data collection.

From the people (P) aspect of STS, we use Shirish et al.'s (2023) work on MB owner-managers to gain insight into the growth and technology mindset of the warung owner. We also use the work of Verhoef et al. (2021) to develop questions related to technology mindset. Regarding tasks, we asked what processes changed before and after becoming SRC partners (Blumberg et al., 2019), particularly related to the procurement of goods.

The questions related to technology aim to gain insights into the technology provided by SRC, whether there have been changes in the features, what actions the companies take when there are changes in the feature and what obstacles occur in the application. Finally, questions about structure attempt to gain insights into how family members help warung owners in their DT process. So, we ask whether the owner has family or relatives that help them run the warung, what kind of activities these family members help the owner with, how often and whether there are formal agreements and rewards for the family member's contribution to the warung operation.

We also asked questions regarding the warung's performance. We refer to Mandviwalla and Flanagan (2021), who indicate that DT values are operational efficiency (reduced cost), effectiveness (increased profit), and additional sales to sustain operations. We ask whether DT had any impact on operational efficiency in terms of transportation costs and whether the warung experienced increased profits (Nwankpa & Roumani, 2016). Finally, we also asked about the surrounding community where the warung operates because it influences the success of the warung business.

4 Data Analysis

Three warungs are involved in this study. Information about the owners is shown in Table 1. We collected data through semi-structured interviews in Bahasa Indonesia, which were transcribed and translated. We undertook two or three interviews with each warung owner; each interview lasted 30–60 min. The interviews were conducted onsite so the researcher could observe the operation and the type of SRC applications used by the owners. We also obtained data by direct observation where we asked the warung owners to try to perform three common tasks in the SRC application, namely, input sales of goods, procurement of goods, and accounting. The researchers documented how the warung owners performed these tasks. If the owner could conduct the tasks without much problem, we classified it as successful. In this way, the observations validate the owner's statement regarding their technology efficacy in the interviews. Data analysis is conducted by reading the interview transcript line by line and labeling the word chunks that represent certain meanings as codes.

Table 1. Owner's Demographic.

Respondent	Owner (Gender)	Owner Age (Years)	Prior occupation	Warung locations
Warung 1	Female	43	Housewife	Close to university campus
Warung 2	Male	41	Trader	In front of apartment
Warung 3	Male	60	Retiree	Close to apartment

5 Findings

The findings from three SRC warung owners are summarized in Table 2. To answer the research question on how the lived experiences of warung owners are affected by DT, we organize and discuss our findings based on socio-technical factors and environment.

5.1 Growth Mindset to Manage Business Environment.

The growth mindset of SRC warung owners, which reflects their ability to transform and create new opportunities by utilizing digital technology in warung operations, is the key to their success. A warung owner with a high growth mindset has the openness to learn, can solve business problems and invent ways to provide value to the customers. A growth mindset also enables the warung owner to overcome the effect of the environment (location and market demand). In our findings, W1 has a high growth mindset as the owner can seize opportunities from DT. W1, which used to be just a traditional warung that did not utilize technology at all, is now able to digitize by joining as an SRC partner. W1 is actively involved in adapting to the new technology, completed all the missions from SRC and even got involved in the SRC community. The lived experience of W1 is thus remarkable and central to the success of the warung.

W2 and W3, on the other hand, are categorized as mid-growth mindset. Unlike W1, who has the motivation and capability to grow with the SRC DT program, W2 and W3 have less apparent growth. Initially, W2 opened a photocopying business and joined SRC in 2008 because the owner was interested in the benefits offered. However, W2 struggled to adapt to the SRC DT program launched in 2019. The combined effect of the COVID-19 pandemic coupled with the lack of support from the SRC agents negatively affected W2's adaptation efforts.

W2 admitted, "It (warung) used to be neater than this, and there were more items. But COVID-19 hit…Now, I still find it difficult to adapt again after COVID. The products we have in the warung are not as good as before, and the customers who come here are also not as many as before."

Table 2. Summary of Findings from SRC Warung Partners

Warung (W) Owner	Indicators								Observation
	People (Growth Mindset)	People (Technology Mindset)	Task	Technology	Structure	Environment	Increase profit	Operational Efficiency	
W1	High Growth Mindset	Proactive Mindset	V	V	V	V	V	V	**Success**
W2	Middle Growth Mindset	Reactive Mindset	X	•	•	•	•	V	**Limited**
W3	Middle Growth Mindset	Reactive Mindset	•	•	V	V	•	V	**Challenged**

Legend
V = Successful • = Partly successful X = Unsuccessful

One of the reasons for the lack of success appears to be the warung owner's inability to move on from the original idea of SRC providing business assistance toward DT. As expressed by W2 "in the past, the SRC provided capital to us, which I felt very helpful. There was a lot of development provided by SRC to its partners, so in the past it was more fun and easier. But because now it is the era of technology and now SRC has also changed the system, everything is more towards the application, so there is no development given to all partners."

Like W2, W3 also has a mid-growth mindset. W3 states "Previously, when we completed the mission, we received points which could be cashed in and transferred through the bank at the end of the year. This increased my profits. Now, I can obtain some gains, but it is unstable: in some months it goes up but in others it goes down". The lived experiences of W2 and W3 are thus less positive than for W1. We observed that W3 in particular had a more dependent attitude towards SRC; when her expectations were not met, this negatively affected the lived experience.

5.2 Technology Mindset – Task and Technology

The level of technology mindset of warung owners represents their ability, self-determination, and willingness to learn the SRC technology. All three of these relate to their lived experiences. The level of technology mindset of SRC Partners indicate the extent to which they can utilize and integrate this technology (the three mobile applications) independently or with assistance from SRC into their operational tasks and activities.

W1 demonstrates a Proactive Technology Mindset by successfully adapting to SRC technology. These adaptation skills are evident through W1's ability to understand SRC technology acquired from training and asking the SRC agent. The high motivation and enthusiasm of SRC partners is evident in their efforts to learn SRC technology gradually, as well as taking advantage of all available learning opportunities. W1 also shows high self-determination in operating the applications provided by SRC. Significant differences were felt by SRC partners before and after using SRC technology. These are also supported by our observations as W1 can carry out all tasks in the three applications AyoToko, AyoKasir, and PojokBayar. She can clearly explain all the features in each application. She also actively finishes all the missions in AyoToko. Her lived experience as a warung manager is correspondingly more satisfying.

However, in W2 and W3 the situation is different. It is true that W2 and W3 can learn and adapt to the technology provided by SRC through interaction with SRC agents. However, they have not fully utilized the learning opportunities, they appear to be more technically dependent on others, and thus they cannot perform all the tasks using the technology provided by SRC. For these reasons, we classify both as having a Reactive Technology Mindset. This reactive technology mindset is also reflected in their sub-optimal lived experiences.

W2 takes longer to learn about the technology from SRC. As explained by W2 "The transition of the move is difficult, I never used an application like this. The process of learning to use this application is also quite long. But now I use this application." Now, W2 only uses AyoToko and PojokBayar, but not AyoKasir. In AyoToko, W2's owner can explain well how he undertakes the process of ordering goods. But W2's owner only orders two items regularly. He does not order other items because they do not sell well. W2's owner rarely completes SRC's missions. He does not use and cannot explain well the function of the other menu items. For the PojokBayar, he can explain how to order credit and other functions, but he does not use it because he feels that the price is more expensive. Based on the interview and observation, we classify W2 as unsuccessful in conducting their tasks and limited in their technology adoption.

Like W2, W3 also only uses two applications (AyoToko and PojokBayar). Interestingly, the owner of W3 stated that "There is no difficulty when using the application. The only problem is when there are errors in the application, although it very rarely happened". W3 can explain clearly and perform how to order goods smoothly using AyoToko. W3 always completes missions and can explain them well. Unlike W2, W3 can explain well the PojokBayar application because he ordered credit, tokens, and data packages from the application. W3 does not use AyoKasir because he claims he has no space to put the cash machine. Another reason is that W3 feels that he is not technologically savvy and does not feel the need to use the cashier application. Based on the interview and observation, we classify W3 as partly successful in conducting their tasks and challenged in their technology adoption.

5.3 Structure in Supporting the Task and Technology Adoption.

The structure turns out to significantly impact the success of a warung. W1 is a prominent example because the owner has strong support from her family. The owner of W1 was able to manage the distribution of tasks well so that the warung operational process continued to run smoothly. Positive interaction between W1 and other SRC partners is a key factor in supporting the DT process. The support received from fellow warung owners under Paguyuban (SRC community) has a positive impact on adaptation to technological change. The establishment of Paguyuban reflects SRC's commitment in providing assistance and resources to their partners to ensure mutual success.

W1 realizes that collaboration and interaction with other owners has significant added value in warung development. This awareness creates a mutually supportive work environment and enables the exchange of experience and knowledge between partners. Furthermore, the high level of support provided by the SRC agent to W1 is an important pillar in improving the quality of warung operational processes. With high quality interactions between W1 and other SRC partners, as well as from SRC, the DT and development of warungs can take place better and more efficiently.

Unlike W1, W2 is only partly successful when it comes to structure. W2 faces several challenges that can affect operational efficiency. There is an uneven division of labor within W2, where almost all warung processes are carried out by one person. This condition can put additional stress on one individual and affect the overall performance of the warung. In addition, W2's interaction with the SRC Partner community and SRC agents appears to be less active. W2 is rarely involved in communication through the WhatsApp (WA) platform and SRC, resulting in a lack of knowledge and experience exchange with other owners. Support from SRC agents is also limited. Infrequent visits of SRC agents to W2 also result in a lack of access to the latest information related to SRC.

W3 has a high level of support from the family in running the warung. W3 has a clear division of tasks between the husband and wife owner team, so the operation of the warung runs well. In addition, W3 has active interactions with other SRC partners. They have received assistance from other SRC partners, which contributed to improving the quality of W3. This positive collaboration creates a mutually beneficial relationship between SRC partners. Not only that, W3 also showed good communication with SRC, where SRC agents regularly make visits to W3. This ensures that W3 is always able to get useful information from the SRC side, strengthening their involvement in the SRC partner network.

5.4 Operational Efficiency

In summary, all three warung owners admit that SRC's DT has helped them achieve operational efficiency.

W1 claimed "Yes, buying goods from SRC is like a one-stop shop. You buy there, and you can get everything. No need for middlemen. If you buy outside, we move from one store to another, we must queue, and the price is different. We must be observant of which one offers a cheaper price." W2 shared a similar feeling and noted the potential to save the transportation costs "At SRC, we order and wait for the goods. If we order in large quantity, we can ask the goods to be delivered so there are savings in transportation costs. The advantage for us is also we do not have to queue." Finally, W3 explained "For the purchase of goods, the price is the same. But now shopping with their system has become more practical. Regarding transportation, there are savings because shopping above a certain price will be delivered by SRC here. So, it saves more on transportation costs and saves time as well. If we buy outside, it is uncertain, and we must wait to buy goods."

6 Discussion

The findings allow us to identify how the livelihoods of warung owners are influenced by DT. All warung owners admitted that the SRC DT program helps them achieve high operational efficiency, which is the key to a significant increase in profits. However, not all warung owners increased their profit, implying that other factors play a role in driving the success of warung DT.

Our study contributes to the literature in several ways. First, the research broadens the existing literature on DT in MBs by integrating it with STS theory in the context of Indonesia. We confirm Shirish et al.'s (2023) finding that the growth and technology mindset of MBs owners plays an important role in their DT success. We found that the growth mindset influenced the warung owner's ability to strategically address their business challenges. A warung owner must have the business capability in managing inventory, attracting customer, to ensure the profitability and sustainability of a warung. An additional advantage comes from a good understanding of the strategic location of warungs.

Secondly, our study provides a special context as MBs in Indonesia have low technology efficacy. Previous work by Mandviwalla & Flanagan (2021) focused on MB owners who are very aware of what others have called the convergence of SMAC (social, mobile, analytics, and cloud) (Khurana et al., 2022). We found that warung owners do not have such efficacy. SRC provides mobile technology that can be used by SRC partners to support the operations of the warung. However, we found that the owners' technology mindset significantly determines their ability to adapt and use the technology to do their tasks differently, which is important in a warung's success. When warung owners can undertake all the tasks and use the technology provided by SRC, their operations will be more effective and efficient. On the other hand, if warung owners cannot handle their tasks well or cannot adopt the SRC technology, then this can lead to serious impacts on operations and hinder future growth.

Additional efforts and support are needed from family, friends, or community to help warung owners overcome these obstacles, so that they can achieve maximum success. A warung owner who gets high levels of structural support from their families and SRC and has a high level of technology adoption can increase their profit. On the other hand, warung owners who do not have support from their family and SRC cannot successfully embrace the technology. The community provided by SRC can also create a positive impact on the implementation of tasks & technology. Community fosters enhanced interaction among partners, offering a platform for sharing knowledge, collaborating on problem-solving, and gaining a deeper understanding of effective technology utilization. Good discussions and clear knowledge transfer within the community play an important role in stimulating efficient use of technology, thus creating optimal conditions for SRC partners to manage their warungs successfully. Ultimately, the lived experience of the warung owners also depends on the extent to which they are successful in adapting to the new technology.

The findings align with Mkansi (2022), who observed that MBs adoption of e-business relies on cohesive collaborative interpersonal teams or ecosystems. They emphasized that these interpersonal ecosystems provide various solutions for MBs including financial assistance, knowledge exchange, workforce enhancement, and primary revenue streams. The study also noted a dependence on familial connections or family members within small e-business retail firms for financing, marketing, and staffing tasks. MBs will depend on external ecosystems to facilitate their transformation. Mandviwalla and Flanagan (2021) stressed the importance of accessing expert ecosystems as drivers for DT in micro-businesses. Mkansi (2022) highlighted the significant roles of non-governmental organizations (NGOs) and government entities in supporting MSMEs.

7 Conclusion

We discuss how DT affects MBs in the form of warung in the Indonesian context. The use of technology and the warung's performance were explored in the context of the support received from SRC. We find that the level of performance following DT varies markedly. The findings of our study are relevant to the SRC, as well as other platforms that create MB-focused DT initiatives, in driving their DT program. Considering the typically low levels of technology efficacy in the warung owner population, SRC needs to ensure that their agents can fully support them. In the long term, they need to design technology from the perspective of users with low technology efficacy levels. They also need to leverage the owner's community to ensure continuous knowledge transfer between successful and challenged partners. This will help to enhance the lived experience of the warung owners and correspondingly help them further to be ambassadors for the SRC programme.

Our study is limited by the small number of cases in a specific region. Future studies should extend this work by investigating other platforms in other geographical areas in Indonesia and other locations in the Global South. More in-depth investigations are needed to gain a deeper understanding of the challenges that MB owners face when operating the technology in the context of a DT initiative to propose more concrete solutions.

Disclosure of Interests. All authors have no competing interests to declare that are relevant to the content of this article.

References

Bai, C., Quayson, M., Sarkis, J.: COVID-19 pandemic digitization lessons for sustainable development of micro-and small- enterprises. Sustain. Prod. Consumption **27**, 1989–2001 (2021). https://doi.org/10.1016/j.spc.2021.04.035

Behera, R.R.: Indonesia EB2B – Reviving the 'Warungs' in a Post COVID world I Part 1. Redseer strategy consultant (2020). https://redseer.com/newsletters/indonesia-eb2b-reviving-the-warungs-in-a-post-covid-world-part-1/

Blumberg, M., Cater-steel, A., Rajaeian, M.M., Soar, J.: Effective organisational change to achieve successful ITIL implementation Lessons learned from a multiple case study of large Australian firms. J. Enterp. Inf. Manag. **32**(3), 496–516 (2019). https://doi.org/10.1108/JEIM-06-2018-0117

Surabaya, H.: Dukung Transformasi Digital UKM, Telkomsel gelar roadshow Digital Creative Entrepreneurs (DCE) di Surabaya, 24 January 2024. https://hariansurabaya.com/2024/01/24/dukung-transformasi-digital-ukm-telkomsel-gelar-roadshow-digital-creative-entrepreneurs-dce-di-surabaya/

Hermawan, D.: Challenge of culture-based SMEs digitalization, 22 August 2022. https://www.bi.go.id/en/bi-institute/BI-Epsilon/Pages/Tantangan-Digitalis asi-UMKM-Berbasis-Budaya.aspx#:~:text=Based%20on%20data%20issued%20b y,of%20total%20investment%20in%20Indonesia

Indonesia Investment: Micro, small & medium enterprises in Indonesia: backbone of the Indonesian economy, 16 July 2022. https://www.indonesia-investments.com/finance/financial-col umns/micro-small-medium-enterprises-in-indonesia-backbone-of-the-indonesian-economy/item9532

Isnurhadi, R.: Empowering warung with digital solutions (2022). https://www.xendit.co/en/blog/empowering-warung-with-digital-solutions/

Khurana, I., Dutta, D.K., Singh, A.: SMEs and digital transformation during a crisis: the emergence of resilience as a second-order dynamic capability in an entrepreneurial ecosystem. J. Bus. Res. **150**, 623–641 (2022). https://doi.org/10.1016/j.jbusres.2022.06.048

Mandviwalla, M., Flanagan, R.: Small business digital transformation in the context of the pandemic. Eur. J. Inf. Syst. **30**(4), 359–375 (2021). https://doi.org/10.1080/0960085X.2021.189 1004

Mkansi, M.: E-business adoption costs and strategies for retail micro businesses. Electron. Commer. Res. **22**(4), 1153–1193 (2022). https://doi.org/10.1007/s10660-020-09448-7

Mukharomah, N., Putra, R., Lenggana, W.F.: Sampoerna retail community program corporate social responsibility communication strategy in Marga Mulya, Bekasi City. Formos. J. Sustain. Res. **2**, 2659–2676 (2023). https://doi.org/10.55927/fjsr.v2i11.6861

Nwankpa, J.K., Roumani, Y.: IT capability and digital transformation: a firm performance perspective. In: Proceedings of the 37th International Conference on Information Systems, pp. 1–16 (2016)

OECD: Enterprises by business size. OECD DATA (2023). https://data.oecd.org/entrepreneur/ent erprises-by-business-size.htm#indicator-chart

Priyono, A., Moin, A., Putri, V.N.A.O.: Identifying digital transformation paths in the business model of SMEs during the COVID-19 pandemic. J. Open Innov. Technol. Market Complex. **6**(4), 1–22 (2020). https://doi.org/10.3390/joitmc6040104

Shirish, A., Srivastava, S.C., Panteli, N.: Management and sustenance of digital transformations in the Irish microbusiness sector: examining the key role of microbusiness owner-manager. Eur. J. Inf. Syst. **32**(3), 409–433 (2023). https://doi.org/10.1080/0960085X.2023.2166431

Verhoef, P.C., et al.: Digital transformation: a multidisciplinary reflection and research agenda. J. Bus. Res. **122**, 889–901 (2021). https://doi.org/10.1016/j.jbusres.2019.09.022

Vide, R.K., Hunjet, A., Kozina, G.: Enhancing sustainable business by SMEs' digitalization. J. Strateg. Innov. Sustain. **17**(1), 13–22 (2022)

Wessel, L., Baiyere, A., Ologeanu-Taddei, R., Cha, J.: Unpacking the difference between digital transformation and IT-enabled organizational transformation. J. Assoc. Inf. Syst. **22**(1), 102–129 (2021). https://doi.org/10.17705/1jais.00655

Yasa, A.: Sampoerna Retail Community hadirkan inovasi digital bagi UKM (2024). https://eko nomi.bisnis.com/read/20190511/12/921414/sampoerna-retail-community-hadirkan-inovasi-digital-bagi-ukm

French-Style Applications of Artificial Intelligence to Human Health in a European Context

Dominique Desbois[✉]

Paris-Saclay Applied Economics, INRAE-AgroParis Tech, 22 Place de L'Agronomie, CS 20040,
91 123 Palaiseau Cedex, France
dominique.desbois@agroparistech.fr

Abstract. In this paper, we discuss the applications of artificial intelligence to human health and the problems this poses in France. We explore the field of application of weak artificial intelligence to human health problems in relation to questions of socio-economic development, law and ethics. We also look at issues arising from strong Artificial Intelligence. We present a few examples of applications of weak artificial intelligence in various fields: predictive medicine, precision medicine, diagnostic and therapeutic assistance, supportive care, computer-assisted surgery, epidemic prevention and pharmacovigilance, and access to care. Within the European regulatory context of the single digital market and the personal data protection regulation, we illustrate the societal issues surrounding the application of artificial intelligence. We conclude with the prospects for the development of artificial intelligence applications in human health in France within a European context.

Keywords: Artificial Intelligence · Human Health · France · Europe

1 The European Specificities of Artificial Intelligence in the Field of Human Health

Artificial intelligence (AI) systems[1], the definition of which has just been updated by the OECD, are a rapidly expanding field of research and development. Their applications are at the heart of human health issues: computer-assisted operations, remote patient monitoring, intelligent prostheses, personalized treatments, etc. AI offers a wide range of approaches, including natural language processing, ontology construction, data mining, machine learning and other decision-making aids. For the purposes of this paper, we will

---

[1] "An AI system is a machine-based system that, for explicit or implicit objectives, infers, from the input it receives, how to generate outputs such as predictions, content, recommendations, or decisions that can influence physical or virtual environments. Different AI systems vary in their levels of autonomy and adaptiveness after deployment", November 29, 2023: https://www.oecd.org/digital/artificial-intelligence/

© IFIP International Federation for Information Processing 2024
Published by Springer Nature Switzerland AG 2024
R. M. Davison and D. Kreps (Eds.): HCC 2024, IFIP AICT 719, pp. 110–122, 2024.
https://doi.org/10.1007/978-3-031-67535-5_10

distinguish between two types of AI: weak AI, which designs and implements artefacts capable of assisting users with tasks; and strong AI, which is defined as a form of AI that attempts to simulate the cognitive functions of the brain [1]. In this paper, we mainly explore the field of application of weak AI to human health problems in relation to questions of socio-economic development, law and ethics, with a small foray into strong AI.

The development of AI systems (AISs) in the healthcare sector, in contact with patients and practitioners, is giving rise to the societal issues of individual consent, health risk management, data confidentiality, data re-use, protocol validation and all the issues relating to human health.

To address these societal issues, the European Commission (EC) submitted a regulatory proposal in April 2021. This proposal provides for: harmonized rules for the marketing, commissioning and use of AI systems in the EU; a ban on certain AI practices; specific requirements for high-risk AISs and obligations for the operators of these systems; harmonized transparency rules for AISs designed to interact with natural persons, emotion recognition systems and biometric categorization systems, as well as AISs used to generate or manipulate pictorial, audio or video content; specific rules for market monitoring and surveillance.

As a preliminary, let us briefly describe the economic and legal context in which the issues surrounding the development of AISs applied to human health are unfolding, through its two essential components: the Digital Single Market for the economic dimension, and the protection of health data for the legal dimension.

2 The Digital Single Market

The Digital Single Market (DSM), launched in 2015, aims first and foremost to stimulate the supply of services throughout the European Union, via the digital transition towards an inclusive digital society, thanks to the development of standards for interoperability on the Web. In the area of personal services, the MUN should help the countries of the European Union (EU) to meet the challenges of public health by promoting digital access to health data for patients and professionals. In the field of health, the effective operationality of the MUN comes up against a number of barriers, both operational and regulatory, particularly for the transfer of health data between EU countries, which may seem paradoxical.

On 15 December 2020, the European Commission proposed two Digital Single Market regulations: The Digital Services Act and the Digital Markets Act. The deployment of the MaSanté@UE platform enables health data to be shared securely between European professionals. France is one of the first eight Member States to join the scheme.

3 The European Framework for Health Data Protection

The purpose of the Regulation on access to electronic data (EU2018/1807) is to remove these barriers. The stakes are high: according to estimates by the European Commission (EC), the MUN in full operation would represent up to 415 billion euros (€). Announced by the Juncker Commission in May 2015, the MUN is being implemented by the European Commissioner for the Digital Society with the support of Directorate-General CONNECT. Among the measures envisaged are: the abolition of roaming charges; cross-border portability of digital content; a ban on geographic blocking; the strengthening of cybersecurity capabilities; the development of broadband with fifth-generation (5G) mobile phones; and the strengthening of personal data protection, particularly that relating to health.

The General Data Protection Regulation (GDPR), which came into force in 2018, contributes to the European desire to create a MUN in a context of data avalanche (Big data) in order to allow the sharing of information and the free circulation of non-identifying data, within the European Union. In particular, the RGPD (EU2016/679) reaffirms the highly sensitive nature of health data and the need to secure it using appropriate procedures.

4 Significant Technological Development in Healthcare AI

Since the 1980s, the symbolic approach in AI, based on logical rules, has enabled expert systems to be developed: for example, Mycin to identify bacterial infections or Sphinx to detect jaundice. Thanks to the increased power of computers and the greater relevance of programming languages, current decision support systems are more effective: for example, the European Desiree project[2], based on a specific ontology, uses the symbolic approach to help practitioners in the treatment and monitoring of breast cancer.

The numerical approach, based on calculations using observed data, has also benefited from the increase in computing power, the development of new algorithms or improvements to existing ones, and the development of storage and addressing capacities enabling the creation of gigantic repertoires of specific data. The mathematical representation of a multi-layer neural network has considerably improved the performance of machine learning algorithms: deep learning algorithms can now detect melanoma from photos of a patient's skin, or detect diabetic retinopathy. These performances are based on self-learning carried out on very large samples: 50,000 photos for melanomas, and 128,000 fundus for retinopathies.

In their ability to produce texts deemed relevant in a matter of seconds, the performance of generative AI systems (SIAgen) is impressive, but sometimes at the cost of approximations, biases and even errors [2]. Their potential applications in the healthcare sector are vast, ranging from assistance with writing information notes, theses or projects, to research programs. The recent development of conversational agents such as ChatGPT, based on large linguistic models (LLM), has improved the performance of tools for extracting information from clinical reports. According to [3], GPT 3.5's performance in selectively sorting publications for inclusion in a literature review is superior to that of a single researcher and comparable to that of two researchers working independently in duplicate. However, the use of conversational agents for complex medical tasks such as establishing a diagnosis or selecting a treatment protocol has highlighted the current limitations of these systems: the more complex the clinical description and

---

[2] Inserm Unit 1142 at Limics, Medical Informatics and e-Health Knowledge Engineering Laboratory (www.limics.fr).

the more precise the question, the more their errors multiply. ChatGPT cannot be used to make medical decisions. Further research is being carried out on enriching the corpus of clinical observations and creating LLMs specific to the medical field, with specific protocols for formulating queries.

Another area of AI is robotics, which aims to develop the autonomy of artefacts by giving them enhanced capacities for perception, decision-making and action, and is seeing its applications develop and extend their field of action: computer-assisted surgery, intelligent prostheses, and robots to assist people with motor or cognitive disabilities.

Research into AI applied to healthcare aims to increase the technical capabilities of these systems in line with medical practices, to provide real added value for patients and doctors that justifies the financial and environmental costs involved in their design and distribution. The design of systems that are transparent to the user and adapted to care pathways therefore remains a priority, as does the structuring and anonymization of highly heterogeneous health data.

5 Health Data Feeding Artificial Intelligence

According to the World Health Organization (WHO), the use of AI systems in the health-care sector requires the following issues to be taken into account with regard to the data used: confidentiality of individual information, security and integrity, documentation of sources, transparency of use, auditing and risk management, compliance with existing regulations in healthcare and related areas, particularly those developed by the European Union[3].

The methodology for collecting and the quality of the documentation of training and test samples is the bottleneck of digital medical AI systems. The data accessible is not initially collected with the aim of developing this kind of software. France has a national medico-administrative data system (SNDS), bringing together descriptions of pathologies, hospital procedures and pharmaceutical prescriptions. However, the use of this data is tricky, as it was collected for the purposes of economic analysis of healthcare services and not for specific medical analyses. Quality controls have revealed an error rate of up to 30% in the description of pathologies. Furthermore, there is not necessarily a connection with the patient's file, so the treatment of a respiratory problem does not necessarily mention the cancer affecting the patient. The confidentiality and appropriate use of this individual health data are protected in France by the 2016 Law for a Digital Republic and the 2018 European General Data Protection Regulation. Accessible only to research projects validated by ethics committees, the processing of this anonymized individual data is subject to the informed consent of the people concerned.

---

[3] For example, Eudralex for the pharmaceutical sector, GMP for manufactured goods, and GDP for the goods distribution sector, not forgetting the RGPD for the protection of personal data.

6   A Few Examples of the Application of Weak Artificial Intelligence to Human Health

The aim of the SUOG (Smart Ultrasound in Obstetrics & Gynecology) system[4] is to improve the real-time interpretation of ultrasound examinations during pregnancy by offering the operator, confronted with unusual parameters, iterative advice and ultrasound images validated by experts during the examination. This system incorporates a specific ontology linking anatomical structures and technical parameters to the various types of pathological pregnancy, enabling the sonographer to approach the level of expertise of the best sonographers. Developed by LIMICS in collaboration with the Trousseau Hospital, the SUOG-based OPPIO project[5] aims to provide decision support in ultrasound analysis for the detection of ectopic pregnancies.

Combining symbolic approaches with supervised statistical learning algorithms, the Epifractal system[6] aims to detect patients at high risk of osteoporosis-related fractures, using automatic natural language processing to incorporate them into specific care protocols. However, systems based on supervised learning, while effective, require further research before they are sufficiently reliable for widespread use.

In psychiatry, the PsyCare project aims to detect cases of psychosis at an early stage. Recent research has shown that the earlier schizophrenia and chronic psychosis affecting adolescents and young adults are treated, the more effective the treatment. In addition, the project aims to target therapy in the early phase on the basis of identified biomarkers in order to propose a therapeutic strategy that can be adapted to each individual case.

One strategic field of application is the interpretation of medical images to help diagnose pathologies, monitor patients and prepare surgical procedures. This involves combining patient-specific information with physiological knowledge models that can be represented by ontologies or graphs. The France Genomic Medicine 2025 plan[7] envisages bringing together all these different individual patient data (genomic, clinical, biological, imaging) on a dedicated platform so as to be able to identify therapeutic profiles enabling better targeted treatments to be developed with a more favorable success rate for pathologies such as cancer, diabetes and even certain rarer diseases.

In the field of textual analysis, unsupervised learning algorithms (with no prior learning phase based on a sample) are also used to discover structural relationships and unidentified categories of interest from the processing of very large volumes of data. In this way, it is hoped to identify new risk factors, personalize treatments, assess their effectiveness, improve pharmacovigilance and predict certain epidemics.

The design of such systems, whether for decision support, diagnosis, risk forecasting or therapeutic design, requires a renewal of the interfacing methodologies between the artefact and the human, as well as a major investment in interdisciplinary dialogue between the various public health experts.

---

[4] SUOG project – https://www.suog.org/suog/.

[5] Towards ontology-based decision support systems for complex ultrasound diagnosis in obstetrics and gynecology - PubMed (nih.gov).

[6] EPIFRACTAL | Health Data Hub (health-data-hub.fr).

[7] Plan France Médecine Génomique 2025 – PFMG 2025 (aviesan.fr).

Thus, thanks to the development of multilayer neural systems and large linguistic models, weak AI is on the way to establishing itself as a new decision-making tool in the diagnosis and prevention of health risks. In view of the possible moral implications of these applications, it is necessary to examine the societal issues surrounding the application of weak artificial intelligence to human health, in order to study the ethical principles that are likely to apply.

7   The Prospects for Strong AI in Understanding the Human Brain and Its Pathologies

The concept of strong AI can be thought of as a theoretical form of AI that attempts to simulate the cognitive functions of the human brain. The Human Brain Project (HBP) is one of a number of research projects that are close to the AI stronghold. Between 2013 and 2023, more than 150 academic institutions from 19 European countries will be involved, thanks to €607 million in EU Flagship funding. It has produced more than 3,000 academic publications and more than 160 digital tools, available on the EBRAIN open science collaborative platform[8]. The highlights of HBP research include the most detailed three-dimensional (3D) atlases of the brain in the world, advances in individualized brain medicine and the development of new technologies inspired by the way the brain works in the field of artificial intelligence (neuromorphic computing). This project is helping to improve our understanding of how the brain works and to better pinpoint the causes of certain brain diseases (Alzheimer's, Parkinson's and Charcot's). A number of researchers are trying to gain a better understanding of how neurons work and how their connections behave, so as to be able to mimic the brain's cognitive abilities. A positioning article details the prospects for digital neuroscience research over the next decade[9].

8   The Societal Issues Surrounding the Application of Weak Artificial Intelligence to Human Health

For these decision support systems to be integrated into the daily practice of each specialist, it is necessary to design their man/machine interfaces in such a way as not to impose a cognitive overload through inappropriate alerts because they do not really apply to the specific clinical context of the patient being treated. In addition, the recommendations proposed by the decision support system must be understood by the practitioner if they are to be integrated, and therefore explicitly justifiable. For this reason, symbolic approaches based on the application of traceable predictive rules appear to be more appropriate than numerical approaches where the algorithm based on the optimization of a quantitative criterion by a multilayer neural network often constitutes in fact a "black box" for the end user.

---

[8] EBRAINS: Europe's Research Infrastructure for Brain Research -.https://zenodo.org/records/10035197.

[9] The coming decade of digital brain research - A vision for neuroscience at the intersection of technology and computing (zenodo.org).

On the other hand, learning databases are likely to suffer from a number of biases resulting from preconceptions made at the time of their creation or the creation of logical inferences, or even during the design of the algorithms. The most common biases are the over-representation (or, conversely, the under-representation) of a category of the population, the categorization often being geographical or ethnic, and sometimes genetic. The difficulty is compounded by the existence of a bias-variance dilemma in unsupervised learning: bias can arise from a lack of generality in the learning sample, while variance results from sampling fluctuations. A bias-variance trade-off is necessary because it is generally not possible to minimize both sources of error simultaneously. Boosting techniques combine biased models to reduce their bias, while bagging techniques combine the least biased models to reduce their variance. At present, finding an optimal technical compromise between performance and interpretability, depending on the context of application, remains the domain of AI research.

The reproduction, or even amplification by reinforcement, of biases not detected a priori in learning databases makes AI a factor in potential discrimination, especially as these so-called "cognitive" biases are difficult to detect a priori: this Achilles heel of algorithms is highlighted in France by a recent report from the Parliamentary Office for the Evaluation of Scientific and Technological Choices (OPECST) [4]. Specialized ethics committees, such as the Commission de réflexion sur l'éthique de la recherche en sciences et technologies du numérique, have been set up to guard against possible abuses resulting from the inappropriate use of AI systems. The experts' ethical recommendations are also incorporated into professional codes of ethics concerning artificial intelligence, such as the Code of Ethics and Professional Conduct of the International Federation for Information Processing [5]. Unesco Recommendation on the Ethics of Artificial Intelligence was adopted in November 2021 by its 193 member states.

In fact, the best way to protect against cognitive bias in the design and development of AI systems is to force AI operators to open up their "black boxes" to public and independent expertise, with the aim of striking a lasting balance between technological development and the protection of citizens. In France, the Villani's report [6] proposes developing the "auditability of AI systems" by recommending "the creation of a body of sworn public experts able to audit algorithms and databases and carry out tests by any means required". Some independent institutions, such as CNIL, are already responsible for this public and independent audit within their remit, although it is still appropriate to strengthen the scientific and technical expertise required in the health sector. Better integration of protective legislative provisions into professional practices and harmonization of health regulations at European level is one way forward, through the development of European standards specific to health AIs.

While public health remains a national competence of the Member States, Article 168 of the Treaty on the Functioning of the European Union (TFEU) gives the European Union additional competences that go well beyond crisis management: improving public health; health information and education; preventing illness and diseases, and threats to physical and mental health; combating the major health scourges, by promoting research into their causes, their transmission and their prevention; monitoring, warning of and combating serious cross-border threats to health; reducing drugs-related harm. In addition, Article 173 of the TFEU stipulates that the European Union shall share competence with the Member States for encouraging better exploitation of the industrial potential of policies of innovation, research and technological development. Provided that there is a common political will to address public health needs, these provisions of the TFEU provide a framework for enhanced cooperation between Member States in the field of technological innovation in the service of public health, particularly in the field of artificial intelligence, where France, through the research programmes it has already developed, can provide leadership and coordination for certain European projects in its specific areas of competence.

Moreover, in many areas of international law, such as the protection of personal data, the European Union is a significant standard-setter at international level [7], if not the dominant standard-setter on its own territory, as demonstrated by the case law of the Court of Justice of the European Union (CJEU) and the adoption of the General Data Protection Regulation (GDPR 2016/679).

The Artificial Intelligence Act (AI Act) was adopted by the European Parliament in March 2024 and promulgated on 21 May 2024 by the Council of the European Union to regulate AI systems in order to encourage their development in Europe while limiting any possible abuses or perverse effects. Among its legal provisions, specific rules for generative AI will apply to verify the quality of the data used in the development of algorithms and compliance with copyright. Artificially generated sounds, images and text will have to be identified to prevent attempts to manipulate opinion. The AI Act will be fully applicable 24 months after its entry into force, but some parts will apply with shorter deadlines: six months for the ban on AI systems presenting unacceptable risks; nine months for the codes of good practice; 12 months for the rules on general-purpose AI systems having to comply with transparency requirements. The rare bans concern applications that run counter to European values, such as citizen rating, predictive policing based on profiling, mass surveillance, or categorisation according to sexual, religious and racial criteria.

Although this regulation is not specific to AI systems in the healthcare sector, many of its provisions are likely to have a drastic impact on areas of application in the life sciences, particularly in human health. These include the pharmaceutical industry, which is undergoing an unprecedented process of digitisation, from pre-clinical research and drug development to patient monitoring and medical workflow management. Medical devices (including specific software) and in vitro diagnostic tools are explicitly taken into account and regulated by the AI Act, having a major impact on the medical technology industry. Lastly, even if they are not yet subject to the CE marking requirements applicable to medical devices, many suppliers or users of advanced digital technologies could see their healthcare applications qualified as low- or even high-risk AI systems.

9   Specific Features of French Human Health Law

Concurrently, the French law of 24 July 2019 on the organization and transformation of the healthcare system (OTSS) strengthened the protection of health data. The French legislator also created a new specific advisory committee: the Comité éthique et scientifique pour les recherches, les études et les évaluations dans le domaine de la santé (CESREES), with enhanced prerogatives. CESREES assesses the scientific and methodological quality of research projects requiring the use of individual health data, as well as their public interest (objectives, expected benefits, transparency, data integrity, scientific quality, etc.).

For example, the PMSI or Programme de Médicalisation des Systèmes d'Information (Information Systems Medicalisation Programme) houses a database containing all the information relating to stays, consultations and treatment procedures carried out in health establishments. This database contains administrative and medical information relating to any patient who has been followed up in a healthcare establishment. It is highly sensitive to the risk of disclosure of confidential health information because, although individual information is anonymous, individuals can be re-identified by cross-referencing with public directories on the basis of geographical and socio-demographic criteria. Following requests by the company SEBDO, publisher of the magazine "Le Point", submitted between 2018 and 2020 to process PMSI data in order to draw up its annual ranking of hospitals and clinics, the Commission national Informatique et Libertés (CNIL), a civil authority independent of the government, had given its authorization for access following opinions issued by its Advisory Committee and subsequent improvements to the ranking methodology. However, in an opinion issued on 2 June 2022, CESREES pointed out other methodological limitations in establishing the ranking, with regard to the indicators used and their weighting. On the basis of the information presented, it considered that "the construction of the indicators used in the ranking may lead to the dissemination of erroneous information on the actual relative performance of healthcare establishments, which could mislead patients and therefore be contrary to the public interest". Once again referred to by the CNIL for further analysis, the CESREES reaffirmed, on the basis of the missions entrusted to it by the OTSS Act and the prerogatives granted to it, the absence of any public interest in processing presenting "major methodological biases". However, while the CNIL refused access to the PMSI database, it emphasized that Le Point magazine remains free to use other sources as part of its journalistic activities (questionnaires, public data, interviews, etc.).

10   Correcting Inequalities in Access to Healthcare: Towards a European Public Health Area

In the developed Western economies, territorial inequality in access to medical care remains one of the major obstacles to the effectiveness of public health policies [8] without ignoring the situation in the global South, which is sometimes much worse.. In France, major disparities in the supply of healthcare services persist, whether at regional level along a North-South gradient or at sub-regional level between urban centers and rural areas, and also, within urban areas, between advantaged and disadvantaged neighborhoods [9]. Despite an increase in the life expectancy at birth indicator between 2002 and 2015, Europe is also affected by these territorial inequalities, with a major divide between East and West, covering disparities in economic development.

On 3 May 2022, the European Commission published a draft regulation for the European Health Data Space (EHDS)[10], which aims to regulate the use of health data within the European Union. In France, the Agence du numérique en santé (ANS) is the

---

[10] https://health.ec.europa.eu/ehealth-digital-health-and-care/european-health-data-space_en.

national contact point for MaSanté@UE, the European cross-border e-health program. The ANS has opened the European e-Health Service (SESALI), which provides secure, structured access to a European patient's medical summary. This service, co-funded by the European Union, implements the recommendations of the European directive aimed at ensuring continuity of care for European citizens, beyond the borders of their country of residence, in compliance with the provisions of the RGPD.

11 Outlook: Proven European Expertise but a Synergy Dynamic to Be Confirmed

According to the Pipame forecast for the French Ministry of the Economy, the artificial intelligence market in healthcare was worth €6.1 billion in 2021. According to converging private sources, the market is estimated to be worth €10.7 billion in 2024, with a forecast of €33.9 billion by 2026[11].

As far as weak AI applied to human health is concerned, future prospects lie technically in the combination of symbolic approaches based on the construction of ontologies specific to health problems and numerical approaches based on the learning capabilities of algorithms specialized in the mining of medical texts of various kinds, following the example of the PsyCARE project.

However, the economic balance sheet for French healthcare AI has to be confirmed: France has undeniable intellectual resources (5,300 researchers in 268 teams) that are being courted by international laboratories, thanks to decisive scientific and technical innovations. However, this know-how has not yet translated into the emergence of companies capable of establishing European leadership in one of the most promising application sectors for AI, according to demographic projections, due to the ageing of the population.

At the VivaTech 2024 trade fair for European innovation, French President Emmanuel Macron announced a €400 million investment plan to finance the training of AI specialists in centres of excellence, as well as a 25% investment by the French government in an investment fund to support sectors technologically linked to AI. BpiFrance has announced a €200 million investment in the French joint venture H, dedicated to AI. According to Agence France Presse, France and the United Arab Emirates have signed a strategic partnership for AI.

For AI investments taken in isolation, the macroeconomic impact remains difficult to estimate because we do not have sufficient hindsight. However, according to some microeconomic studies, AI is likely to have a positive impact on productivity in certain sectors: PwC, for example, has calculated that labour productivity is almost 5 times higher in the sectors most exposed to AI. It should be remembered, however, that despite the equally thunderous announcements of the time, the contribution of information and communication technologies to growth is finally estimated for the OECD area, depending on the country considered, at between 0.2% and 0.5% for the first half of the 1990s, then between 0.3% and 0.9% during the second half of the 1990s[10].

---

[11] These figures are given for guidance only, as estimates vary widely from one source to another, due to the assumptions and estimation methodologies used.

A key societal issue for the effectiveness of weak AI approaches is the ability of public authorities, whether national or European, to articulate their ambitions in terms of human health with a regulatory policy based on controlled access for medical research to health data considered to be 'common goods' and the inclusive conditions for making them available in order to obtain informed consent from patients. According to the Villani report, to achieve this, public authorities must initiate new modes of data production, collaboration and governance, through the creation of "data commons". The State, as a trusted third party, can play a major role by taking measures to encourage the various players involved in public policy on human health to share and pool their individual data.

Questions remain about both the dynamics of economic investment in AI in the various fields of application of health and the political will of the Member States of the European Union (EU). Under the Treaty on the Functioning of the European Union (TFEU), the EU's competence in the area of health is limited to providing support to its Member States [11]. By way of illustration, the multiannual budget devoted by the EU to health is relatively small: €449.4 million for 2014–2020, compared with the €1.6 billion in public funding for French AI research and the €80 billion devoted by the European Union over the same period.

Synergies between EU Member States are therefore needed at the level of fundamental and applied research to promote investments likely to transform methodological advances in AI into technological innovations that protect or improve public health. Also, the recent establishment of various health agencies (European Food Safety Agency, European Medicines Agency, European Centre for Disease Prevention and Control) is likely to broaden the scope of European competences (respectively, health risk assessment, marketing of food or medicines, fight against infectious diseases). Joint decision 1082/2013/EU of the European Parliament and of the Council makes it possible to organize the European Union's response to serious cross-border threats, such as the Covid 19 crisis.

However, other EU instruments make it possible to intervene more indirectly in the area of public health: the application of the principle of free movement of goods to medical devices and medicinal products (the European Medicines Agency assesses the risks associated with placing a medicinal product on the market and the European Union defines the technical characteristics that medical devices placed on the market must meet); the free movement of healthcare professionals and patients organized by European regulations (Directive 2005/36/EC, revised in 2013, sets out the conditions for automatic mutual recognition of healthcare qualifications acquired in another Member State, and EU Directive 2011/24 of 2011 on the application of patients' rights in cross-border healthcare aims to guarantee patient mobility).

12 A Privileged Position for France in the European Space for Healthcare AI

AI is playing a central role in the current therapeutic revolution, and France currently occupies a privileged position in Europe in this field. First of all, France has an extremely favorable ecosystem for healthcare AI, with top-level research teams, a world-renowned network of university hospitals, European, if not global, pharmaceutical and high-tech companies, and a cluster of innovative healthcare AI joint ventures. Promoting the improvement of the healthcare system for the long-term benefit of patients, the National Health Strategy adopted by the French government encourages the pooling of health data, promoting the creation of empirical databases that can support unbiased learning processes. Through independent institutions such as the CNIL, scientific ethics committees such as Comets, legal provisions such as the Research Code and the Public Health Code, its research funding instruments and universal healthcare cover, France offers a research and innovation environment that is safe from any potential commercial, ideological or villainous abuses in the field of human health. In terms of potential progress, a number of French medical teams are already able to analyze complexes of health data to guide therapeutic strategies, particularly for cancer treatments where the design of innovative molecules depends on the molecular anomalies detected in tumors. Healthcare AI thus helps to target and evaluate the resulting therapeutic strategies, optimizing patient access to these innovative but costly therapies.

13   Disclosure of Interests.

The author has benefited from the logistical support of the CREIS-Terminal association and of its research institute during this work of synthesis and reflection. The author has no other competing interests to declare that are relevant to the content of this article.

Acknowledgments. The author would like thank the anonymous reviewers for meaningful comments and suggestions. He alone remains responsible for any errors or omissions and for the opinions expressed in this document.

References

1. Pennachin, C., Goertzel, B.: Contemporary approaches to artificial general intelligence. In: Artificial General Intelligence, Cognitive Technologies, pp. 1–30. Springer, Heidelberg (2007). https://doi.org/10.1007/978-3-540-68677-4_1
2. Académie Nationale de Médecine (2024). *Systèmes d'IA générative en santé: Enjeux et perspectives*, 5 March
3. Viet-Thi, T., et al.: Sensitivity and specificity of using GPT-3.5 turbo models for title and abstract screening in systematic reviews and meta-analyses. Ann. Intern. Med. (2024). https://doi.org/10.7326/M23-3389
4. de Gannay, C., Gillot, D.: Information Report # 464, Office parlementaire d'évaluation des choix scientifiques et technologiques (2017)
5. Kreps, D.: IFIP code of ethics. In: Goedicke, M., Neuhold, E., Rannenberg, K. (eds.) Advancing Research in Information and Communication Technology. IAICT, vol. 600, pp. 405–420. Springer, Cham (2021). https://doi.org/10.1007/978-3-030-81701-5_17
6. Villani, C., et al.: Donner un sens à l'intelligence artificielle, pour une stratégie nationale et européenne. Prime Minister Office, French Government, Paris (2018). ISBN: 978-2-11-145708-9
7. Cohen-Tanugi, L.: L'Europe comme puissance normative internationale: état des lieux et perspectives. Revue Européenne de Droit **2021/2**(3), 100–106 (2021)

8. Bärnighausen, T., Bloom, D.E., Humair, S.: Going horizontal: shifts in funding of global health interventions. N. Engl. J. Med. **364**(23), 2181–2183 (2011)
9. Vigneron, E.: Inégalités de santé, inégalités de soins dans les territoires français. Les Tribunes de la Santé **38**, 41–53 (2013)
10. Colecchia, A., Schreyer, P.: ICT investment and economic growth in the 1990s: is the United States a unique case? A comparative study of nine OECD countries. Rev. Econ. Dyn. **5**(2), 408–442 (2002)
11. Gruny, P., Harribey, L.: Information Report # 648, European Affairs Commission, French Senate (2017)

Survey Instruments for Measuring Digital Inequality at the Individual Level

Ahmed Imran$^{(\boxtimes)}$ ⓘ, Marjia Haque ⓘ, and Farhan Shahriar ⓘ

University of Canberra, Canberra, ACT 2617, Australia
{ahmed.imran,marjia.haque,farhan.shahriar}@canberra.edu.au

Abstract. Fresh light has been shone on the problem of digital inequality (DI) in recent years due to its expanding complexity and impact on society. An issue with DI research is the need for more reliable instruments to measure and gauge the DI status quo, which is critical for appropriate remedial measures and solutions. DI has several dimensions beyond traditional concepts, such as uneven access to modern technologies, insufficient digital literacy, and limited internet use. Measures of DI often ignore some subtle and hidden elements, including socio-economic status, digital literacy, and skill, giving an imprecise image of the differences among people. This paper utilises existing literature, builds on a theoretical framework developed earlier, and subsequently explores possible survey instruments to measure DI. The paper also emphasises the need for constant improvement of survey tools due to their dynamic nature, influenced by rapid technological changes, to capture the complexities of digital inclusion. The proposed instruments are a starting point that will fill a void and allow researchers in the area to conduct more empirical studies to test and validate the suitability of the instruments to better understand the phenomena for valid comparisons and multiple applications.

Keywords: Digital Inequality · Digital Divide · Digital inclusion · Digital Inequality Index · Survey Instrument · Measuring Inequality

1 Introduction

Digital inequality (DI) is a pressing social concern that profoundly impacts our society. With the increasing digitalisation of almost all our communication and services, including healthcare, education, and civic engagement, DI is increasingly seen as a factor that can worsen existing inequalities and bring about new ones. People who do not have adequate access to digital resources are at a disadvantage while society undergoes a significant transformation. Consequently, DI worsens social cohesiveness, restricts educational opportunities, increases economic inequality, and jeopardizes public health [1]. Research has already established its growing implications across a broad range of domains [2, 3]. As such, addressing these digital inequalities is crucial for building sustainable digital societies through enhancement of relevant models with new variables, theories, and impacts [4].

© IFIP International Federation for Information Processing 2024
Published by Springer Nature Switzerland AG 2024
R. M. Davison and D. Kreps (Eds.): HCC 2024, IFIP AICT 719, pp. 123–134, 2024.
https://doi.org/10.1007/978-3-031-67535-5_11

For over two decades, researchers and legislators have expressed concerns regarding the disparities faced by individuals with uneven access to and benefits from technology. Comprehensive knowledge and conceptualization of the DI and its expanding implications for society are becoming increasingly necessary in the current era [3]. The concept of "digital inequality" has progressed beyond the "digital divide" (DD), which was seen as the availability and competency of individuals with information technology compared with those without it [5]. DI is now broadly defined as a situation "where groups differing in characteristics such as socioeconomic background, age, and gender are disadvantaged in terms of access, knowledge, competency, and costs with respect to digital resources" [1].

Because of its complexity, measuring and comprehending DI has become more complex. Most of these DD assessments primarily examine the degree of access to ICT among various demographic categories, often contrasting more technologically sophisticated groups (such as younger and more educated individuals) with less technologically advanced groups (like the elderly and those with low incomes) [6].

While scholars have come a long way in understanding how intricate and multidimensional DI is, there still needs to be a widely used framework or set of measures for measuring it. The investigation of DI across various areas, people, and historical periods is challenging due to the lack of standardized measurement instruments [3, 4, 7]. Tackling emerging facets of DI requires researchers and practitioners to exhibit flexibility in adapting their measurement methodologies and keeping pace with the evolution of technological innovations. DI has evolved into a more complex phenomenon, with variations in the material, cultural, cognitive, and social resources needed to get the most out of the technology [8–10]. Advancements in technology, particularly in artificial intelligence, machine learning, and the Internet of Things ((IoT), have compounded the progression of DI, leading to disproportionate value and outcomes amongst different recipient groups/people within the society [11].

Addressing DI issues successfully with its multifaceted nature will pose a challenge for policymakers, educators, and practitioners unless there is a consistent and shared vocabulary for assessment. It has become imperative to develop strategies that effectively bridge gaps and advance digital inclusion for everyone to create a resilient and harmonious community. The positioning of this research within the context of the specific phenomenon of DI is thus crucial for laying the foundation to shape DI theory and to address a critical practical problem that society is grappling with as emphasized by Chatterjee and Davison [12]. With that motivation, the paper aims to develop instruments to measure DI and determine the Digital Inequality Index (DII) at an individual level.

2 Method

This research is part of a multi-phase, multi method [13, 14] wider research aiming towards a theoretical framework of DI.

This paper is a follow up phase of the previous phase of this research which identified the major factors influencing DI through an exhaustive scoping review of the existing body of knowledge following five stages [15]. The review advanced the work of DiMaggio and Hargittai [16] by offering a conceptual framework on multiple dimensions of

DI, which was carried out almost two decades ago with no significant work in between. This conceptual framework for factors influencing DI [1] served as the foundation for this paper to develop possible instruments and to measure the DII at individual level.

This phase of the research is dedicated to developing instruments and providing potential measurement instruments and models for assessing DII. This was accomplished by following these steps:

Step 1. It was envisaged, one of the most crucial initial steps towards a viable measurement model is to define the construct well within the contextual domain, where indicators should include the whole domain [17]. While the review from the previous phase provided the basis for this phase, a further literature search was carried out using specific keywords to define and augment the description of each construct, followed by a critical examination of established instruments to measure those constructs.

Step 2. Survey measurement items are classified into two broad categories: contextual and individual, along with an independent construct, purpose, based on the classifications of constructs [1]. A slight modification of the constructs, including the addition of a new factor, was introduced following further review and analysis of the factors.

Step 3. The weighting of factors is determined by a panel of experts based on their relative importance after a thorough discussion and brainstorming session among the participants. The panel comprise a group of academics from diverse disciplines with varying experiences, the majority of whom are members of the University's DI research cluster. They included two senior IS professors, one senior professor specializing in mental health, one professor from the Business and Economics department, two IS academics (Associate Professors/Senior Lecturers), and four junior academics/research assistants.

Step 4. Finally, a simple formula was developed to determine individual DII scores, based on survey responses.

3   Past Attempts to Measure DI

The historical method of measuring DI follows the development of digital technology acceptance, use, and access. Assessing the accessibility of fundamental infrastructure, including internet connectivity and computer access, was the focus of DI measurement in the early phases. As technology developed, the focus moved to comprehending skill gaps and digital literacy. This historical viewpoint acknowledges socioeconomic class, geography, and demographics' role in determining digital disparity. Researchers have created indices and measures throughout time that consider age, education level, and income.

A few attempts have been made to define DI so that it can be operationalized to measure DI in some form, with a limited focus mainly on accessibility and use. Table 1 presents those initiatives and their indicators aiming to measure DII at a group or community level and not at individual level.

While the above initiatives offer some perspectives and insights based on different contexts, the instruments used to measure DI are predominantly focused on accessibility, connectivity, or usage, overlooking many other factors that directly contribute to an individual's DI status [1].

Table 1. Past attempts to measure DII

Measure	Indicators (Weight)	Focus	References
Digital Divide Index (DDIX)	• Percentage of computer users (30%) • Percentage of computer users at home (20%) • Percentage of Internet users (30%) • Percentage of internet users at home (20%)	Social groups of EU member countries, assumed to be disadvantaged; Used Eurobarometer public opinion survey	[18]
Digital Divide Index (DDI)	Made up of two scores, the socioeconomic (SE) score (income, gender, age, and education) and the infrastructure/adoption (INFA), both have a value between 0 and 100, with 100 being the greatest digital disparity • Percentage of computer users 50% • Percentage of Internet users 30% • Percentage of internet users at home 20%	Four potentially disadvantaged categories in Slovenia (Women, People aged 50 years and over, low-education group, low-income group)	[7]
Digital Inclusion Index (DII)	Considers variables, such as the cost of technology, digital skills, digital device availability, and internet connectivity	Selected disadvantaged groups in Hong Kong: (Older people; New arrivals; Single parents; Female homemakers; Young people aged 10–17 in low-income households; People with a disability or chronic illness)	[6]
Australian Digital Inclusion Index (ADII)	Measures digital inclusion along the three aspects of accessibility, affordability, and digital ability using survey data	Data from the Australian Internet Usage Survey, conducted between June and December 2022, served as the foundation for the 2023 study. Then 2020 and 2021	[40]

4 The New Multiple Factors Influencing DI

DI is a complex construct consisting of social, behavioural, technical, and legislative components, each influencing the phenomenon to varying degrees. Hence, a good understanding of the characteristics of its multidimensional components is critical. Imran et al. [1] identified a complex set of individual and contextual factors through an extensive review of contemporary literature that can practically influence DI and its consequences [1]. Nevertheless, a novel construct, 'Health & IQ' has been introduced in response to recent studies highlighting the substantial impact of health and psychological well-being on DI among individuals [19, 20]. This addition aims to encompass and address all possible factors in the assessment of DII. Figure 1 captures the dimensions of DI leading to the index of the concept at an individual level.

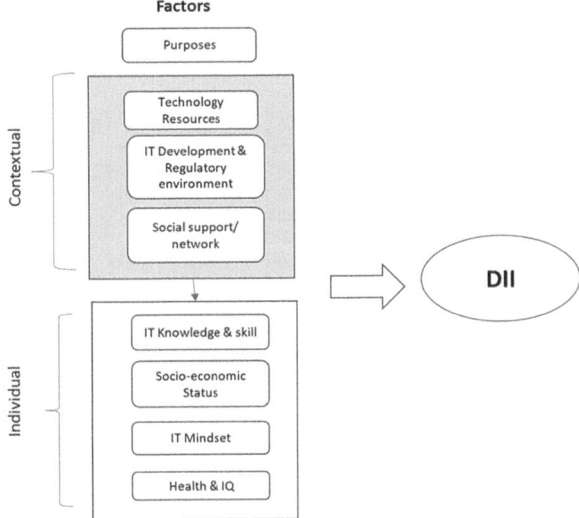

Fig. 1. Factors influencing DI-adapted from Imran, et al. [1]

Factors are divided into two major categories: contextual and Individual. Each of the categories consists of four factors. The contextual factors include technological resources, regulatory environment, and social support/network. The individual factors include IT knowledge and skill, socio-economic demographics, personal IT behaviour and health/IQ. The following section provides the details of these constructs.

5 Proposed Constructs/variables with Measurement Items

Constructs represent latent variables, primarily formative and reflective, based on their underlying nature and how they are measured [21]. In reflective constructs, Indicators are seen as interchangeable and reflective of the underlying construct, as such changes in the latent variable cause changes in the observed variable [22]. Formative models/constructs are used when the observed variables together on the combination of indicators define the latent construct [23], which is, in this case, a composite index of DI, providing an overall score of DI.

Configurational or Structural equation modelling (SEM) could be used to gauge the degree of influence of various factors and the synergistic effects between and among various factors contributing to DI. SEM allow a combination of both reflective and formative constructs, with careful consideration of their theoretical foundations, measurement models, and estimation and interpretation procedures [24]. Configurational models explore complex causal relationships to understand how different combinations of factors contribute to various outcomes in real-world settings [25, 26].

Based on a previous study [1], Table 2 enumerates the proposed measurement instruments for DI collected from established studies and bodies. Some constructs have been renamed and expanded to enhance clarity and to facilitate the incorporation of more relevant indicators to measure DII.

Table 2. Survey measurement items to measure DII

Variables	Definition	Measurement items	Proposed weight	Studies
Purposes (Pur)	What is the reason/purpose for using technology?[27], p140	**Pur1:** Social purposes (connecting with friends and families) **Pur2:** Informational purposes (to get information using Google) **Pur3:** Instrumental purposes: (online banking, shopping, bill payment and other services)	(15%)	[27]
Contextual				
Technology resources (TR)	Custody or authorization to use digital technologies	**TR1:** Access to Desktop, Laptop, Printer etc **TR2:** Access to internet and/or wi-fi use **TR3:** Access to Smartphone, iPad/Tablet **TR4:** Affordability for using ICT devices	(15%)	[28]

(continued)

Table 2. (*continued*)

Variables	Definition	Measurement items	Proposed weight	Studies
IT development and regulatory environment (ITDRE)	This construct involves indicators that reflect the ease, support, and infrastructure the government provides in utilizing ICT services in the country where the individual resides. The construct can be best measured by two indexes developed by the ITU	**IDI (The ICT Development Index):** A composite index/assessment of a country's ICT development, combining three dimensions: access, usage, and skills **IRT (The ICT Regulatory Tracker):** A robust data-based tool for benchmarking and the identification of trends in ICT legal and regulatory frameworks of a country, which records the existence of relevant regulatory frameworks and features	(10%)	[29, 30]
Social support/network (ScNet)	Relationships between individuals [31]	**ScNet1:** Friends/families' support to learn online services **ScNet2:** Colleagues' support in IT tools, and techniques **ScNet3:** Community's support to use IT services	(10%)	[32, 33]
Individual				
Knowledge/ skill (ITK)	'The acquisition of awareness or facts, data, information, ideas, or principles to which one has access through formal or individual study, research, observation, experience, or intuition' [34] p 451	**ITK1:** Operational Internet skill (connect Wi-Fi, download apps etc.) **ITK2:** Informational Internet Skills (to get info) **ITK2:** Communicational Internet Skills (communicate through the Internet) **ITK3:** Mobile Internet skill (use mobile and internet) **ITK4:** language skills	(15%)	[35]
Socio-economic Status (SES) (Sc)	A measure of one's combined economic and social status [36]	**SES1:** Income Level **SES2:** Education level **SES3:** Homeownership **SES4:** Self-reported financial status **SES5:** Occupational Status **SES6:** Job Stability **SES7:** Subjective (perceived) social status (social standing) within the country of residence **SES8:** Perceived status within the community where the individual is living	(15%)	[16, 37–41]

(*continued*)

Table 2. (*continued*)

Variables	Definition	Measurement items	Proposed weight	Studies
IT Mindset (ITM)	IT mindset is an individual character trait that implies a position of the mind, including a deeply embedded viewpoint, belief, interest, and behaviour towards IT [42] This construct consists of two sub-constructs: IT beliefs (ITB): A person's subjective judgements of IT [43] Personal Innovativeness of IT (PIIT): The willingness of an individual to try out any new information technology [44]	**ITB1:** IT is going to make my job easier **ITB2:** Our existing processes/working mode needs to be changed through IT **ITB3:** IT adoption will benefit me personally **PI1:** If I hear about a new IT, I look for ways to try it **PI2:** Among my friends and colleagues, I'm usually the first to try out new IT **PI3**: I like to experiment with new IT	(10%)	[42–44]
Health/ IQ (HIQ)	Computer anxiety: Fears about the implications of computer use such as the loss of important data or fear of other possible mistakes [45] Cognitive abilities: Brain-based skills needed to conduct tasks	**HIQ1:** Physical health conditions (vision/hearing problem, any physical illness, disability) **HIQ2:** Computer anxiety **HIQ3:** Cognitive abilities/ psychological issues	(10%)	[19, 20]
Total Weight			100%	

The weight of the factors, as shown against each variable, was determined by a panel of experts, indicating their relative importance in defining the formative construct [21]. The suggested weights are expected to undergo additional refinement, influenced by specific circumstances, and subjected to further validation through empirical studies. The points were totaled to determine the scores for the individual. Each indicator will be measured through 1–7 Likert scale with appropriate questionnaire. While each of the demographic indicators such as age, gender, nationality, and ethnicity could independently influence DI status, they have been excluded from the calculation of DII. This decision stems from their independent and binary nature of the variable, ensuring that DI index remains a universal measure applicable to all citizens, free form demographic biases. The demographic information will, however, be collected as part of the survey. This data will be used for cross-comparison of DII status quo amongst various groups and entities.

The proposed DII of an individual can be calculated as follows:

$$DII = Pur \times 0.15 + (TR \times 0.15 + ITDRE \times 0.10 + ScNet \times 0.10) + (ITK \times 0.15 + SES \times 0.15 + ITM \times 0.10 + HIQ \times 0.10).$$

6 Validity and Reliability

While the chosen indicators are conceptually relevant and have empirical support from previous research, testing the model's fit, including the validity and reliability of constructs through statistical indices, is crucial to measure DI precisely.

The robustness of the model can be evaluated through sensitivity analysis - examining how changes in the measurement model impact the overall model [46]. Expert opinions and feedback on the formative constructs are also an accepted practice, where experts provide valuable insights into the appropriateness of the chosen indicators, validate theoretical alignment and appropriate weightages of each construct.

Various validity tests through statistical measures like factor loadings, average variance extracted (AVE), and composite reliability ensure that the measurement model accurately captures the reflective constructs and that the selected indicators are valid representations of the underlying latent variable [47]. Finally, the overall fit of the measurement model can be evaluated using the Comparative Fit Index (CFI) or Tucker-Lewis Index (TLI) [48].

The validity of the formative construct can be endured through content validity (assessed through expert judgment, literature review, or other methods to confirm that the chosen indicators represent the full scope of the construct), convergent validity (the extent to which the indicators measure the same construct), and discriminant validity (the extent to which the formative construct is distinct from other constructs) [49].

Reliability is ensured through the observed indicators consistently and accurately measuring the underlying latent variable, which can be carried out through composite reliability or Cronbach's alpha (a higher Cronbach's alpha (typically above 0.70) suggests better internal consistency. The item-total correlations help identify any items that may not contribute sufficiently to the construct's overall reliability [50]. The reliability of each indicator in terms of its contribution can be assessed by examining the significance and strength of the relationships between each indicator and the formative construct (DII). [51].

7 Conclusion

The DI is becoming a pressing problem while technology is developing in rapid scale, affecting communities and individuals on many levels. Developing a reliable set of survey instruments is essential for measuring and comprehending the complex aspects of DI and DI index (DII). Since survey instruments are so extensive, researchers can investigate not just the access and affordability, but also the subtleties around digital inequalities, including the contextual issues that lead to inequality. With these measurement tools, policymakers and relevant stakeholders may be informed about which populations, areas, or demographics may be disproportionately impacted by DI, what benchmark need to be achieved at a particular context/ location and what resources need to be allocated.

The paper has proposed a comprehensive and holistic model with a complete set of instruments. The proposed model and instruments, derived from theoretical insight, represent an initial phase, which were selected after a thorough analysis of relevant studies with proven success. This is expected to undergo further modification and refinement through a discussion and debate, which will then be tested through an empirical study for validation and further enhancement. The ultimate objective is to develop a robust measurement model that can be applied to diverse global context.

Validating the construct requires a careful and rigorous approach, and it may involve iterative refinement of the model based on empirical results and theoretical considerations. It is essential to validate the proposed model on an independent dataset to ensure the stability and generalizability of the measurement model [47].

However, it is also essential to understand the limitations of survey instruments. The dynamic nature of digital technology and its rapid advancements pose a challenge in keeping survey instruments current and reflective of the evolving digital landscape.

Hence, the constructs' stability over time should also be assessed as the same construct is likely to change at different points in time under varying conditions. Furthermore, using self-reported data has some inherent biases. Thus, survey findings should be interpreted cautiously. Nevertheless, survey instruments are valuable tools when combined with other qualitative methods to examine the phenomenon. Stakeholders may collaborate to create focused plans and policies for appropriate IT services and guarantee that everyone can benefit from the digital age by using the insights obtained through these measurement tools. It is expected DI researchers will utilise and advance this work to help addressing this pressing problem more objectively and effectively.

References

1. Imran, A., Quimno, V., Gregor, S.: Factors influencing digital inequality: a scoping review (2022)
2. Robinson, L., et al.: Digital inequalities and why they matter. Inf. Commun. Soc. **18**(5), 569–582 (2015)
3. Imran, A.: Why addressing digital inequality should be a priority. Electr. J. Inf. Syst. Develop. Countries **89**(3), e12255 (2023)
4. Vassilakopoulou, P., Hustad, E.: Bridging digital divides: a literature review and research agenda for information systems research. Inf. Syst. Front. **25**(3), 955–969 (2023)
5. Dewan, S., Ganley, D., Kraemer, K.L.: Complementarities in the diffusion of personal computers and the Internet: implications for the global digital divide. Inf. Syst. Res. **21**(4), 925–940 (2010)
6. Wong, Y.C., et al.: Digital divide and social inclusion: policy challenge for social development in Hong Kong and South Korea. J. Asian Public Policy **3**(1), 37–52 (2010)
7. Vehovar, V., et al.: Methodological challenges of digital divide measurements. Inf. Soc. **22**(5), 279–290 (2006)
8. DiMaggio, P., Hargittai, E.: From the 'digital divide' to 'digital inequality': studying Internet use as penetration increases (2001)
9. Graham, R., Smith, D.T.: Internet as digital practice: examining differences in African American Internet usage. Fut. Internet **3**(3), 185–203 (2011)
10. Hsieh, J.P.-A., Rai, A., Keil, M.: Understanding digital inequality comparing continued use behavioral models of the socio-economically advantaged and disadvantaged. MIS Q. **32**, 97–126 (2008)
11. Heeks, R.: Digital inequality beyond the digital divide: conceptualizing adverse digital incorporation in the global South. Inf. Technol. Dev. **28**(4), 688–704 (2022)

12. Chatterjee, S., Davison, R.M.: The need for compelling problematisation in research: the prevalence of the gap-spotting approach and its limitations. Inf. Syst. J. **31**(2), 227–230 (2021). https://doi.org/10.1111/isj.12316

13. Venkatesh, V., Brown, S.A., Bala, H.: Bridging the qualitative-quantitative divide: guidelines for conducting mixed methods research in information systems. MIS Q. **37**, 21–54 (2013)

14. Venkatesh, V., Brown, S.A., Sullivan, Y.: Guidelines for conducting mixed-methods research: an extension and illustration. J. AIS **17**(7), 435–495 (2016)

15. Grant, M.J., Booth, A.: A typology of reviews: an analysis of 14 review types and associated methodologies. Health Inf. Libr. J. **26**(2), 91–108 (2009)

16. DiMaggio, P., et al.: Digital inequality: from unequal access to differentiated use. In: Social Inequality, pp. 355–400 (2004)

17. Diamantopoulos, A., Winklhofer, H.M.: Index construction with formative indicators: an alternative to scale development. J. Market. Res. **38**(2), 269–277 (2001)

18. Hüsing, T., Selhofer, H.: The digital divide Index-a measure of social inequalities in the adoption of ICT (2002)

19. Micheli, M., Lutz, C., Büchi, M.: Digital footprints: an emerging dimension of digital inequality. J. Inf. Commun. Ethics Soc. **16**(3), 242–251 (2018)

20. van der Zeeuw, A., van Deursen, A.J., Jansen, G.: The orchestrated digital inequalities of the IoT: how vendor lock-in hinders and playfulness creates IoT benefits in every life. New Media Soc., 14614448221138075 (2022)

21. Cenfetelli, R.T., Bassellier, G.: Interpretation of formative measurement in information systems research. MIS Q. **33**, 689–707 (2009)

22. Jarvis, C.B., MacKenzie, S.B., Podsakoff, P.M.: A critical review of construct indicators and measurement model misspecification in marketing and consumer research. J. Consum. Res. **30**(2), 199–218 (2003)

23. Petter, S., Straub, D., Rai, A.: Specifying formative constructs in information systems research. MIS Q. **31**, 623–656 (2007)

24. Henseler, J., Ringle, C.M., Sarstedt, M.: Using partial least squares path modeling in advertising research: basic concepts and recent issues. In: Handbook of Research on International Advertising. Edward Elgar Publishing (2012)

25. Fiss, P.C.: Building better causal theories: a fuzzy set approach to typologies in organization research. Acad. Manage. J. **54**(2), 393–420 (2011)

26. Ragin, C.C.: Redesigning Social Inquiry: Fuzzy Sets and Beyond. University of Chicago Press (2009)

27. Leukel, J., González, J., Riekert, M.: Adoption of machine learning technology for failure prediction in industrial maintenance: a systematic review. J. Manuf. Syst. **61**, 87–96 (2021)

28. Soomro, A.A., Breitenecker, R.J., Shah, S.A.M.: Relation of work-life balance, work-family conflict, and family-work conflict with the employee performance-moderating role of job satisfaction. South Asian J. Bus. Stud. **7**(1), 129–146 (2018)

29. ITU: ICT Development Index 2023. https://www.itu.int/hub/publication/d-ind-ict_mdd-2023-2/. Accessed 06 Feb 2024

30. ITU: Global Digital Regulatory Outlook (2023). https://www.itu.int/pub/D-PREF-BB.REG_OUT01-2023. Accessed 06 Feb 2024

31. Coleman, J.S.: Social capital in the creation of human capital. Am. J. Sociol. **94**, S95–S120 (1988)

32. Kvasny, L.: A conceptual framework for examining digital inequality (2002)

33. Van Deursen, A.J., Van Dijk, J.A.: The first-level digital divide shifts from inequalities in physical access to inequalities in material access. New Media Soc. **21**(2), 354–375 (2019)

34. Wojtczak, A.: Glossary of medical education terms: part 3. Med. Teach. **24**(4), 450–453 (2002)

35. Van Deursen, A.J., Van Dijk, J.A.: Digital Skills: Unlocking the Information Society. Springer, New York (2014). https://doi.org/10.1057/9781137437037
36. Diemer, M.A., et al.: Best practices in conceptualizing and measuring social class in psychological research. Anal. Soc. Issues Pub. Policy 13(1), 77–113 (2013)
37. Ihm, J., Hsieh, Y.P.: The implications of information and communication technology use for the social well-being of older adults. Inf. Commun. Soc. 18(10), 1123–1138 (2015)
38. Hargittai, E., Hinnant, A.: Digital inequality: differences in young adults' use of the Internet. Commun. Res. 35(5), 602–621 (2008)
39. Robotham, D., et al.: Do we still have a digital divide in mental health? A five-year survey follow-up. J. Med. Internet Res. 18(11), e309 (2016)
40. Turrell, G., et al.: Measuring socio-economic position in dietary research: is choice of socio-economic indicator important? Pub. Health Nutr. 6(2), 191–200 (2003)
41. Thomas, M.E., et al.: Separate and unequal: the impact of socioeconomic status, segregation, and the great recession on racial disparities in housing values. Sociol. Race Ethn. 4(2), 229–244 (2018)
42. Imran, A., Gregor, S.: Conceptualising an IT mindset and its relationship to IT knowledge and intention to explore IT in the workplace. Inf. Technol. People 32(6), 1536–1563 (2019)
43. Ajzen, I., Fishbein, M.: A Bayesian analysis of attribution processes. Psychol. Bull. 82(2), 261 (1975)
44. Agarwal, R., Prasad, J.: A conceptual and operational definition of personal innovativeness in the domain of information technology. Inf. Syst. Res. 9(2), 204–215 (1998)
45. Sievert, M., et al.: Investigating computer anxiety in an academic library. Inf. Technol. Libr. 7(3), 243 (1988)
46. Cheung, G.W., Rensvold, R.B.: Evaluating goodness-of-fit indexes for testing measurement invariance. Struct. Equ. Model. 9(2), 233–255 (2002)
47. MacKenzie, S.B., Podsakoff, P.M., Podsakoff, N.P.: Construct measurement and validation procedures in MIS and behavioral research: integrating new and existing techniques. MIS Q. 35, 293–334 (2011)
48. Shi, D., Lee, T., Maydeu-Olivares, A.: Understanding the model size effect on SEM fit indices. Educ. Psychol. Measure. 79(2), 310–334 (2019)
49. MacKenzie, S.B., Podsakoff, P.M., Jarvis, C.B.: The problem of measurement model mis-specification in behavioral and organizational research and some recommended solutions. J. Appl. Psychol. 90(4), 710 (2005)
50. Taber, K.S.: The use of Cronbach's alpha when developing and reporting research instruments in science education. Res. Sci. Educ. 48, 1273–1296 (2018)
51. Lau, A.L., Cummins, R.A., Mcpherson, W.: An investigation into the cross-cultural equivalence of the Personal Wellbeing Index. Soc. Indic. Res. 72, 403–430 (2005)

Author Index

© IFIP International Federation for Information Processing 2024
Published by Springer Nature Switzerland AG 2024
R. M. Davison and D. Kreps (Eds.): HCC 2024, IFIP AICT 719, p. 135, 2024.
https://doi.org/10.1007/978-3-031-67535-5

GPSR Compliance

The European Union's (EU) General Product Safety Regulation (GPSR) is a set of rules that requires consumer products to be safe and our obligations to ensure this.

If you have any concerns about our products, you can contact us on ProductSafety@springernature.com

In case Publisher is established outside the EU, the EU authorized representative is:

Springer Nature Customer Service Center GmbH
Europaplatz 3
69115 Heidelberg, Germany

The manufacturer's authorised representative in the EU is Springer
Nature Customer Service Centre GmbH, Europaplatz 3, 69115 Heidelberg,
Germany. If you have any concerns regarding our products, please
contact ProductSafety@springernature.com

Printed and bound by CPI Group (UK) Ltd, Croydon, CR0 4YY
29/04/2026
02099531-0008